HUMAN GENETICS
THE BASICS

Ricki Lewis

Routledge
Taylor & Francis Group

LONDON AND NEW YORK

First published 2011

by Routledge
2 Park Square, Milton Park, Abingdon, Oxon OX14 4RN

Simultaneously published in the USA and Canada
by Routledge
270 Madison Ave, New York, NY 10016

Routledge is an imprint of the Taylor & Francis Group, an informa business

© 2011 Ricki Lewis

The right of Ricki Lewis to be identified as author of this work has
been asserted by her in accordance with sections 77 and 78 of the
Copyright, Designs and Patents Act 1988.

Typeset in Aldus/Scala Sans
by Florence Production Ltd, Stoodleigh, Devon
Printed and bound in Great Britain
by TJ International Ltd, Padstow, Cornwall

British Library Cataloguing in Publication Data
A catalogue record for this book is available from the British Library

Library of Congress Cataloging in Publication Data
Lewis, Ricki.
Human genetics : the basics / Ricki Lewis.
p. cm.—(The basics)
Includes bibliographical references and index.
1. Human genetics—Popular works.   I. Title.
QH431.L41857 2010
599.93′5—dc22   2010017256

ISBN 13: 978–0–415–57984–1 (hbk)
ISBN 13: 978–0–415–57986–5 (pbk)
ISBN 13: 978–0–203–84058–0 (ebk)

CONTENTS

FIGURES

TABLES

ACKNOWLEDGMENTS

Figures 1.1, 2.3, 2.4, 2.5, 4.2, 4.3, 4.4, 4.5, 5.2, 6.2:
© 2008 *Molecular Biology of the Cell*, 5th edition, by Bruce Alberts et al.
Reproduced by permission of Garland Science/Taylor & Francis LLC.

Figures 2.1, 5.1:
© 2002 *Genetics for Healthcare Professionals*, by Heather Skirton and Christine Patch.
Reproduced by permission of Garland Science/Taylor & Francis LLC.

Figures 2.2, 3.3, 3.4:
© 2003 *Introducing Genetics: From Mendel to Molecule*, by Alison Thomas.
Reproduced by permission of Garland Science/Taylor & Francis LLC.

Figures 3.2, 4.1:
© 2005 *Applied Genetics in Healthcare: A Handbook for Specialist Practitioners*, by Heather Skirton, Christine Patch, and Janet Williams.
Reproduced by permission of Garland Science/Taylor & Francis LLC.

Figure 3.1:
© Donna Polski. Reproduced by kind permission.

Figure 6.1:
© Mickie Gelsinger. Reproduced by kind permission.

ABBREVIATIONS

AAV   adeno-associated virus
ADA   adenosine deaminase
AIDS   acquired immune deficiency syndrome
ALD   adrenoleukodystrophy
ALL   acute lymphoblastic leukemia
AML   acute myeloid leukemia
AV   adenovirus
CODIS   Combined DNA Index System
DNA   deoxyribonucleic acid
DOE   Department of Energy
ER   endoplasmic reticulum
FH   familial hypercholesterolemia
GAO   Government Accountability Office
GWAS   genome-wide association study
HD   Huntington's disease
HDL   high-density lipoproteins
HH   hereditary hemochromatosis
HIV   human immunodeficiency virus
HNPCC   hereditary nonpolyposis colon cancer
IVF   *in vitro* fertilization
LCA   Leber congenital amaurosis
LDL   low-density lipoproteins

MLL	mixed lineage leukemia
MRC	Medical Research Council
NIH	National Institutes of Health
OTC	ornithine transcarbamylase
PKU	phenylketonuria
RNA	ribonucleic acid
RPE	retinal pigment epithelium
SCID	severe combined immune deficiency
SNP	single nucleotide polymorphism
TB	tuberculosis
TIGR	The Institute For Genomic Research
VLDL	very-low-density lipoproteins

FROM ANCESTRY
TO DESTINY

The Rosetta Stone is an ancient Egyptian artifact that made it possible to glean meaning from hieroglyphic writings. The content of the etchings on the stone tablet proved not as memorable as the connections of the symbols to each other. The stone, a statement from Ptolemy V set down in 196 BC, held instructions to erect statues in temples and spelled out a particular tax repeal—news of the day, more or less. But the tablet depicted the messages in three languages—two Egyptian and classical Greek. The lasting knowledge was to come in the comparison of the three languages, and in the ability to go from one to another.

Like the Rosetta Stone, the human **genome** is a set of instructions, also for building a temple of sorts—a human body. Two copies of our genome are tightly entwined inside each of our **cells**, which are the building blocks of our bodies. The three languages of the genome are the informational molecules **DNA**, **RNA**, and **protein**, presented in Table 1.1 and Figure 1.1. A cell can transcribe a gene's worth of DNA sequence information into RNA, and then translate the RNA into a specific sequence of protein building blocks. The proteins are what provide our inherited traits and characteristics. Genes vary slightly from person to person, which, with environmental influences, shapes the nuances of our traits. Also like the Rosetta Stone, it is the bigger picture that may be the most valuable. A genome-wide view of ourselves reveals that our DNA sequences are 99⁺ percent alike, uniting humanity. Yet within that other less than 1 percent of our DNA sequences lie the intriguing differences that are responsible for our individuality and our ills, and perhaps the reasons for our conflicts.

The information in a sequence of DNA building blocks (a gene) is transcribed into a molecule of RNA, which is translated into a protein. The protein is responsible for, or contributes to, the trait or illness associated with the gene.

The human genome, again like a language, transcends time. It holds not only the information for how we survive and function in the present, but compelling clues to our pasts, as well as the ability to change (mutate) in the future. While the instructions themselves—the DNA sequences—may directly dictate how we look and feel, standing back we can compare them and see how we are related to others and our place in the living world.

In short, our understanding of the DNA molecule has provided powerful new ways of looking at ourselves through the lens of time.

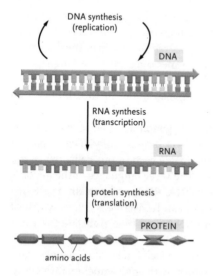

Figure 1.1 From DNA to protein

Source: *Molecular Biology of the Cell*, 5th edition, Figure 1–4, p. 4

Table 1.1 Mini glossary

Cell	The unit of a living organism.
Chromosome	A structure in a cell's nucleus, built of DNA and associated proteins, that contains the genes. A human cell has 23 pairs of chromosomes.
DNA	The genetic material, deoxyribonucleic acid. DNA is a long molecule built of sequences of four nitrogen-containing bases: adenine (A), guanine (G), cytosine (C), and thymine (T). The bases are held in pairs by two backbone structures, forming a double helix.
Gene	A DNA sequence that encodes a protein's amino acid sequence.
Gene pool	All of the gene variants in a population.
Genome	All of the DNA information in the DNA of an organism. A genome is species-specific, and also varies among individuals of the same species to a small extent.
Mutation	A change in a DNA sequence that may be natural or induced.
Protein	A long molecule, built of amino acids, that provides or contributes to an inherited trait.
RNA	Ribonucleic acid is an informational molecule that serves as an intermediate language between DNA and protein.

Consider this first chapter, then, an express time machine, with several stops in the distant past, a quick look at the present, and a glimpse of possible futures.

OUR SHARED DEEP ANCESTRY

We are all, at the core of our genetic selves, African. Anatomically modern humans have lived in Africa for at least 200,000 years, leaving tantalizing but scant clues in the form of bone fragments, footprints, stone tools and evidence of their use. Linguistics provides other clues, because languages followed some of our ancestors as they left Africa, in trickles and surges, to settle the rest of the world.

A DNA CLOCK

Until the 1970s, the sole evidence for the peoples of the past was archeological—shards of a skull, images painted on cave walls, a chipped rock ax. The 1970s brought the ability to determine the sequences of the four building blocks of DNA: **adenine** (A), **guanine** (G), **cytosine** (C), and **thymine** (T). When researchers realized that a DNA sequence **mutates**, or changes, at a known rate, they also realized that the changes in sequence over time could be used as a molecular clock. Following time using a DNA-based clock would lessen reliance on artifacts and remains, for the vestiges of mutations past remain in the genomes of people living today—in all of us.

The basic idea behind a DNA clock is simple and logical. Think of each DNA base in a sequence as a bit of information. The more of a sequence that two present-day people share, the more closely related they are or, put another way, the more likely it is that they shared a recent ancestor. An alternative explanation for genetic similarity is that two individuals have many DNA sequences in common simply by chance. It happens, but rarely.

Having DNA sequences in common is like having anything else in common—it can point to a shared source. Imagine a group of students meeting at their university residence at the start of their first term. All are wearing or carrying something bearing the school name—a cap or notebook—the university insignia a contemporary identifier. As the students chat, four realize that they are from the same region, their accents distinguishing them from the others. Two of those four students are wearing tee-shirts from a bicycle race held in their hometown. After the meeting, they discover that they have been to some of the same music festivals and soccer matches, and had even been scouts together. The two students recognize right away that they are from the same place and, in fact, attended the same primary and secondary schools—they do not share their distinctive accent, attire, and experiences by chance alone.

Genetic ancestry is an area of human genetics that is similar to the two students realizing that they share their hometowns. We seek our similarities, at the DNA level, to reconstruct our origins and identify present-day relatives. Like the university residence scenario, geography is important in making sense of our diversity. Our distribution on the planet has had a profound effect on our genetic ancestry, establishing distinct variations of the human genome in people who remain in the same area for many generations. Such groups that have children only among themselves and retain a unique culture are termed "indigenous." The partitioning of certain gene variants in indigenous populations is dissipating and blurring right now, as human **gene pools** mix as people travel and meet others or are visited by others. (A gene pool is all of the gene variants in a population, not a characteristic of an individual.) Before the echoes from our past fade as genomes homogenize, researchers—genetic archeologists—are comparing the DNA sequences of modern people to deduce where we came from.

Information about our past can be read from how our DNA sequences vary at specific places in the 3.2 billion building blocks of the human genome. Those sequence differences become meaningful when they are plotted as tree diagrams, with new mutations depicted as branchpoints. Once a mutation occurs, it is presumably passed to descendants.

Collections of mutations come to define specific branches, or lineages—family lines. Genetic sleuths have meticulously tracked these inferred branchpoints by comparing DNA sequences among diverse people living today and have reconstructed the major lineages of humankind. The same conceptual diagram can go even farther back in time to show relationships among species.

In 2005, the National Geographic Society began the Human Genographic Project, which is sampling DNA from people all over the world. The project started with indigenous peoples who, by definition, have remained in their homelands for many thousands of years. The portals to the deepest past lie in them, because their unique combinations of DNA sequences have not mixed with those from other groups. However, their numbers are plummeting. Of today's world population of nearly 7 billion, indigenous peoples number only about 380 million. They are scattered, and unfortunately sometimes marginalized, on all continents in many nations (Gracey and King 2009: 65). The Genographic Project has since expanded to include anyone wishing to send in a DNA sample and a low fee, with the goal of tracking at least 100,000 modern human genomes.

Genetic ancestry testing such as that being done in the Genographic Project does not consider entire genomes, but two telltale parts—the Y chromosome, found only in males, and the DNA sequences in mitochondria, which are structures in cells that are passed from women to all of their offspring. (Tests that track other genome parts are increasingly being used.) The Y chromosome and mitochondrial DNA sequences add broad strokes to our portraits of the past, for they reveal only two of the many lineages that contribute to the genome of any one individual. You have four grandparents, but eight great-grandparents, 16 great-great-grandparents, and so on. The farther back you go, the less any one ancestor contributed to your genome.

DNA sequence comparisons usually are consistent with artifact evidence from archeology and fossil evidence from paleontology. Clues from these very different sources converge with our oldest "modern" ancestors, who dwelled in Africa about 170,000 years ago. But the full story begins much earlier, with our closest non-human relatives. For them, although we lack DNA evidence, the few clues that we have provide a context to consider what transpired to make us human. Table 1.2 summarizes major stops on the road to humanity, discussed in the next sections.

Table 1.2 Timeline of humanity

Years ago	Major events
6–7 million	Common ancestor of chimp and humans lived in Africa
5.8 million	*Ardipithecus* dwelled in Africa
2–4 million	Several species of *Australopithecus* lived in Africa
2.5 million	*Homo* emerged in and spread beyond Africa
2 million	Human-like characteristics arose and persisted
200,000	Modern humans in lived Africa; Neandertals left Africa for Europe
170,000	Mitochondrial Eve (most recent common female ancestor)
156,000	Herto people lived in Ethiopia
70,000–100,000	First modern people left Africa but died out
60,000	Y Adam (most recent common male ancestor)
45,000–50,000	People left Africa for southern India, Australia, New Guinea, Middle East
35,000	14 ancestral populations lived in Africa, where Bantu-speakers replaced click-speakers. People migrated from Asia to Europe
20,000	People migrated to Siberia, escaped ice age in areas called "refugia"
5,000–15,000	Agriculture arose and spread

HUMANS EMERGE AND DIVERGE

Between 6 and 7 million years ago, among the dozen or more types of ape that dwelled on the African continent, lived chimp-like animals that were distant ancestors of humans. Over the millennia, mutations happened. Such genetic changes that altered appearance, behavior, or perhaps disease resistance in ways that improved the odds of survival in the particular environment enabled their owners to pass on the helpful gene variants. In this way, over time, the populations of chimp-like animals diversified, perhaps in different ways in different places. We have glimpses of them from fossilized remains at a very special place in Ethiopia, in an area called the Afar Triangle, where the Red Sea, Gulf of Aden, and Main Ethiopian Rifts meet. Since 1981, Tim White, a paleontologist from the University of California, Berkeley, has led teams of students and scientists to the area to meticulously unearth pre-human specimens and the species that lived with them. The area is a mile-deep triangle of roughly 500 kilometers (310 miles) on a side, some 225 kilometers (140 miles) northeast of Addis Ababa, near the Awash River. The sediment layers here reach back nearly 6 million years, providing a slice in time to eyes that can imagine long-ago scenes from the scant and scattered remains.

The oldest remains of human ancestors from the Awash River valley date to 4.4 million years ago, a measurement made possible by radioactive dating of rocks from a nearby volcano. These fossils, of an animal named *Ardipithecus ramidus*, may have been the first **hominin**—animals that gave rise only to us (White et al. 2009: 75). For the past 15 years, a large, international team of researchers has been collecting, cleaning, and trying to make sense of 110 specimens. The fossils come from 36 individuals, and the researchers have so far assembled two partial skeletons, including near-complete skulls, and various bones from the limbs, hands, feet, and pelvis. One was a woman who stood about 120 centimeters tall and weighed about 50 kilograms. The shapes and enamel thickness of the teeth suggest a diet rich in fruit, consistent with the remains of figs and various berries in the area. Her hand and foot bones indicate that she maneuvered around her habitat much differently than a chimpanzee—she didn't walk on her knuckles but stood erect and, rather than swinging through the trees, palmed her way through.

Males and females were about the same size, and the skull was about the same size as that of a chimpanzee.

The fossil find revealed that these pre-people lived in the woodlands and perhaps visited scraps of nearby forest—their fossils were noticeably *not* in the savannah (grasslands), where later hominin fossils were left. The 150,000 other fossils that the researchers collected paint a portrait of a cool, moist woodland that was home to a rich array of birds and small mammals, their delicate bones preserved in owl droppings, as well as the ever-present insect life, and larger mammals that included shrews, rodents, antelopes, elephants, bats, porcupines, and monkeys. Damaged bones testify to the activity of hyenas.

Fossil evidence is more abundant from about 4 million years ago, when ancestors called australopithecines lived. They were a little less chimp-like than *Ardipithecus*, yet clearly more simian than someone you might encounter at the market today. For a few million years, Africa was home to at least five species of these australopithecines, some overlapping in time. Because they likely could not travel far, their populations probably remained isolated and therefore genetically distinct. But the traits that would carve humanity from those long-ago gene pools were already beginning to emerge.

In the australopithecines, an increasingly erect stance made possible life beyond the treetops, on the grasslands. A larger (but still small by human standards) brain fueled the cooperative and cognitive skills that would one day vastly improve nutrition, even permitting the trial-and-error tastings of plants that would become the first medicines. Every once in a while, an accident would preserve some scene from this distant past. A flash flood 3.3 million years ago, in what is now Ethiopia, buried a young girl, leaving intact a skeleton revealing long legs, like those of a human, topped by a stockier upper body reminiscent of a chimp. Her head was larger than that of a chimp, her toes indicated she walked mostly erect, and her teeth were a hodgepodge of those of her ancestors (Johanson and White 1979: 321).

By about 2.5 million years ago, the australopithecines were still hunting and gathering, but some of them had become different enough for scientists to give them their own name—*Homo* (Asfaw et al. 2002: 317). (All of these names are, of course, just our creation.) The transition from *Australopithecus* to *Homo* is murky in the fossil record, but, generally speaking, the newcomers exhibited more sophisticated tool use, family structure, and better control of the environment than the oldtimers. The australopithecines fade from the fossil record as *Homo* debuts and thrives.

Homo erectus fossils are found throughout and beyond Africa. These forerunners were thought to have died out by 100,000 years ago, but recent evidence suggests this might not be the end of their story. In 2004, researchers discovered a complete female skeleton and bones from other individuals beneath a cave floor on the island of Flores in Indonesia. Dating techniques placed her lifetime as about 18,000 years ago. She was about three feet tall, with a brain a third the size of a human's today. Officially named *Homo floresiensis*, the mysterious small people were soon nicknamed Hobbits, after characters in the *Lord of the Rings* trilogy.

Debate continues on exactly who and what the Hobbits were. Early explanations suggested that they were merely small modern people, miniaturized after many generations on an island, where those who could live on less food were more likely to survive—an effect seen in other types of mammal too. Researchers attributed the disproportionately large Hobbit head to a long-ago mutation that persisted. For a time, the Hobbits were thought to descend directly from *Homo erectus*, surviving longer thanks to their isolation. But analysis of the foot bones and relative lengths of limb bones of Hobbit skeletons suggest a more intriguing scenario—their bottom halves were more ape-like than that of Lucy, the little australopithecine girl who was washed away in a flood 3.3 million years ago (Jungers 2010). The Hobbits, then, may be holdovers from an ancestor who predates even the australopithecines!

The Hobbits were likely an aberration, a consequence of their island isolation. Tim White offers a more generalized view, wrapping his vivid imagination around the fossils that festoon his lab and office, envisioning the curious period of *Homo*'s emergence:

> It all started about 2.5 million years ago, when guys started banging rocks together. That allowed the niche to expand, and the start of culture. Tool making, utilizing stone, probably began in *Australopithecus*. These bipeds were small-brained, and they weren't busy becoming human, but being australopithecines. At some point, a population of that highly intelligent, bipedal generalist—*Australopithecus*—began to exhibit behaviors that we

see in the chimps, which hunt monkeys but lack tools. That particular sect of early hominin that didn't lack those tools anymore formed the beginning of the lineage that would ultimately diverge from other australopithecines that kept on being australopithecines, and that lineage would go on to become early *Homo*.

THE FIRST MODERN PEOPLE

About 2 million years ago, some of the qualities that make us human began to appear and, when helpful, persist. Vocalization became language. Walking improved. Stronger necks, larger leg muscles, and bigger butts made us born to run, easing migration and improving diet as hunting skills sharpened. A mutation in the gene encoding a type of muscle protein called "room for thought" shrunk the jaw, enabling the brain to expand (Stedman et al. 2004: 415). Perhaps the gene duplications that today distinguish our genomes from those of our closest non-human relatives appeared at this time, providing a new canvas for the genes that would propel our abilities to remember and learn.

Paleontologists date fossils by referring to their location among rock layers, coupled with chemical clues, such as the timetable of the breakdown of a radioactive isotope into a different element. Genetic genealogists, on the other hand, look from the present back in time, using DNA sequences as their guideposts rather than fossils and footprints. The two approaches are most exciting when they tell the same story. That has happened in the tale of our "deep ancestry" —"deep" compared with traditional genealogy that uses family lore, documents, and photographs, yet shallow when compared with our legacy from an ancestor shared with chimps.

Fossil evidence meets, or at least overlaps, genetic evidence about 160,000 years ago. In 1997, Tim White's team discovered pieces of three human-like skulls sticking out of the dried mud roads of a temporarily abandoned town in Ethiopia, Herto, where floodwaters had chased away the living residents (White et al. 2003: 742). Said White, "What to a nonpaleontologist might have looked like a cow pie, to us was a fossil hominin cranium." It took years to reassemble the skull fragments into something resembling the original heads, and to use potassium–argon dating to arrive at the approximate time of 156,000 years ago. These people, which the researchers named

Homo sapiens idaltu (*idaltu* means "elder" in the local language), had large brains, faces, and teeth, and probably resembled Australian aborigines.

The Herto fossils are important for what they lack as well as for what they have. The skulls were found alone and undamaged—the rest of the bodies were nowhere to be seen. This argues against cannibalism, which would have damaged the skulls, and may even be evidence of mortuary practice or some sort of reverence for the skull minus the torso and limbs, suggesting culture.

The idea of the Herto people being the first, or one of the first, modern humans dovetails nicely with the idea of a "mitochondrial Eve," the term used to denote the most recent female ancestor that all of us living today share. This is where DNA sequences enter the picture.

In the mid 1980s, another Berkeley researcher, graduate student Rebecca Cann, working with mentors Allan Wilson and Mark Stoneking, had the revolutionary idea to use as a clock mitochondrial DNA (mtDNA), which mutates fast (Cann et al. 1987: 31). They would look at differences in the mtDNA sequences of as diverse a group of modern people as they could assemble. As they did so, it quickly became clear that the most diverse present-day human genomes come from Africa. Genomes from elsewhere include parts of that ancestral base set of DNA variants, as well as other mutations that occurred later, marking migratory paths when superimposed on geographical maps.

A mitochondrion is a structure inside cells that houses the chemical reactions that extract energy from food nutrients. Each cell has thousands of them. Mitochondria have their own genetic material, which is what Cann and her colleagues looked to for clues to the past. Because she knew the mutation rate of key mitochondrial genes, Cann could note the DNA sequence differences among African genomes and count backwards. This extrapolation placed the time of emergence of our most recent common maternal ancestor at from 100,000 to 200,000 years ago. Subsequent estimates from another young female researcher, genetic anthropologist Sarah Tishkoff, then at the University of Maryland in College Park, and her colleagues, added mtDNA data from hundreds more modern Africans (Tishkoff et al. 2003: 208). These data refined the date for the approximate start of modern humanity to about 170,000 years ago—an estimate remarkably close to the time of the Herto people, according to their remains.

Clues to the most recent common male ancestor in the Y chromosome sequences of modern males do not reach back as far as the 170,000 or so years for the maternal mitochondrial DNA—only to about 60,000 years ago (Ke et al. 2001: 1151). Why the lag for "Y Adam?" This is where imagination and the nuances of human behavior flesh out the science. An explanation is that, for many years, people mated in small groups, with too few males surviving to leave their Y chromosome mark in modern genomes. Mitochondrial DNA, after all, is passed from a mother to all of her children; a Y chromosome is passed only to males. Perhaps the hunter–gatherer lifestyle led to more male deaths in pursuing a meal than female deaths.

EXPANSION OUT OF AFRICA

Extrapolating back from contemporary DNA sequence diversity, against a backdrop of paleontology and archeology, to trace our ancestors' travels becomes even more compelling when clues from linguistics and geography are added to the mix. The peopling of the planet seems to have been a tango between the potential in our genomes and restrictions in survival imposed by the environment.

People left Africa when the climate cooperated, traversing a temporarily wetter Sahara desert to reach the colder north. We don't know why they went, but an ebb and flow of humanity probably pulsed out of the dark continent, with most surges squelched by any number of environmental challenges. Over the millennia, many small groups probably attempted to leave Africa, dying out along the way (Li et al. 2008: 1100). Hard times in rough climes must have wiped out many a human settlement, enabling only the hardiest, luckiest, or cleverest individuals to survive. Clues in DNA, linguistics, and climatology provide fleeting glimpses of these prehistoric journeys, like peeking at a few frames of a film (Behar 2008: 1130; Liu et al. 2006: 230).

The first people left Africa about 100,000 years ago, but their trails apparently grew cold within 30,000 years. A brutal ice age dawned about 60,000 years ago, but, by 50,000 years ago, as land bridges emerged, some people made it out of Africa. They ultimately colonized southern India (at the time joined to Malaysia and Sri Lanka) and then Australia and New Guinea, which were one land mass. They reached Siberia by 20,000 years ago, establishing the gene pool—more a puddle—that would found the Native Americans, who traversed a land bridge southward over a 5,000-year period (Wang et al. 2007: e185; Goebel et al. 2008: 1497).

When the waters returned, flooding the land bridges over which humanity had journeyed, the distribution of surviving populations made it look as if people had somehow jumped or flown from Africa to Australia, rather than trekking across transient tongues of linking lands. Human skeletons found in New South Wales attest to this colonization, as do distinctive Y chromosome markers (DNA sequences) shared today between people from India and Australian aborigines. Like the students at the university residence meeting, wearing the same striking tee-shirts because they came from the same town, these peoples once shared ancestors in a common homeland.

Australia, India, and Siberia weren't the only places where people expanded. By 50,000–45,000 years ago, others had left Africa, populating the Middle East, where archeological clues pick up the natural narrative. Meanwhile, in eastern Africa, as the forests gave way to savannahs, people spread out over the continent. Sarah Tishkoff's meticulous mitochondrial DNA tracking revealed that, by 35,000 years ago, 14 ancestral populations had become established in Africa. Gradually, Bantu-speaking people from the west began to replace older peoples who spoke "click" languages, such as the Khoisan and Pygmies, isolating them in pockets of the Sudan, South Africa, and Ethiopia.

People began to arrive in Europe from central Asia around 35,000 years ago. Some of them might have been surprised by residents who had already been living there since leaving Africa 200,000 years ago—the Neandertals (Stix 2008: 56; Hall 2008: 34). Their offshoot of the hominin tree would wither away, perhaps because eventually they could no longer survive the brutal cold and didn't have the wits or wherewithal to move away from it. DNA from a preserved Neandertal arm bone has enabled the Neandertal genome to be sequenced, and results so far confirm that they were not our immediate ancestors, but a side branch.

Meanwhile, the ice descending on northern Europe forced the pockets of peoples who *were* our ancestors to move towards the warmer and drier Mediterranean. The areas they moved to are known as "refugia," for they provided escape from the advancing glaciers and cold. By 16,000 years ago, people resumed their westward journeys in climate-driven migrations that paralleled similar episodic journeys in Africa. Again, the migrating Europeans found others already there, descendants of an earlier wave of people who, about 20,000 years ago, had left Africa for the Middle East and the lands ringing the Mediterranean sea. Perhaps they were already sowing the earliest seeds of what would become agriculture.

The observational and reasoning skills required to recognize the value of saving and planting seeds nicely dovetailed with a changing climate favoring grasses. At the same time, small groups who had managed to survive the Ice Age in the Balkans, Iberia, and Italy left other distinctive genetic markers. Today, the men of Europe fall into seven major groups, each defined by a distinctive Y chromosome (Karafet et al. 2008: 830). They represent the legacy of seven men who ventured westward thousands of years ago and were fortunate, or savvy, enough to have left fertile descendants.

Some descendants of long-ago Europeans came to look different from their African forebears, thanks to the ability of DNA to change, and then that of **natural selection** to favor those new variants of physical characteristics that increased the likelihood of leaving fertile offspring. In this way, white people came into being. The first light skin probably arose from a mutation in a gene called *MC1R*, which stands for melanocortin 1 receptor. A receptor is a protein on a cell's surface that functions like a catcher's mitt, binding a particular molecule. When *MC1R* binds such a molecule, it stimulates the cell to produce a dark pigment called eumelanin. If the receptor is not activated, then the cell manufactures a different pigment that produces light skin and hair. As people reached northern latitudes and the generations passed, natural selection favored those with lighter skin (Laleuza-Fox et al. 2007: 1453). A fair complexion let in more ultraviolet radiation from sunlight, needed to manufacture vitamin D. These paler people didn't need the high eumelanin levels necessary in the tropics to protect against the higher exposure to ultraviolet radiation in the stronger sunlight. And so, in the north, people with lighter skin were better suited to their environment.

While in Europe, Asia, and elsewhere the punishing effects of ice ages left their marks on surviving human genomes, people continued to spread throughout Africa. Their genetic diversity reverberates today in the great variety of skin tones, hair textures, facial features, heights, and body forms of modern Africans. In a groundbreaking study, researchers from South Africa, the US, and Australia sequenced the genome of Desmond Tutu, the civil rights activist and a Bantu (the majority population group), and that of a modern-day hunter-gatherer member of the Khoisan people, who live in the Kalahari Desert in Namibia (Schuster et al. 2010: 943). The Khoisan are not only a tiny minority, but are probably the most ancient people, genetically speaking, alive today. The genomes of the two men differed at 1.3 million single-base sites, and none of these sites had the same DNA base as the several dozen other genomes sequenced, mostly from Europeans. Perhaps even more intriguing is that partial genome comparisons showed that the Khoisan man and two others from his group are less alike than is a European to an Asian! The Khoisan, named !Gubi, also had gene variants that reflect natural selection favoring adaptations to desert life (Table 1.3).

The little groups of humanity that emerged from the ice ages began to approach each other, and even coalesce, by about 10,000 years ago, thanks to a new force—agriculture. Hunter-gatherers no longer had to hunt or gather. With a reliably static and consistent food source, populations grew explosively and spread. Communities began to mix and mingle, rather than replace each other. Genetic lines blurred. The rest of the story, as they say, is history.

Table 1.3 Khoisan gene variants that show adaptation to life in the desert

Gene	Adaptation
ACTN3	Sprinting ability
TAS2R38	Ability to taste bitter substances, enabling detection of poisons in plants
CLCNKB	Chloride channel that conserves water
DARC	Absence of variant that protects against malaria, not needed because environment is not wet enough to sustain malaria mosquitoes

OUR RECENT PAST: FROM ROYAL BASTARDS TO LOST COLONIES

Most people who delve into their DNA to illuminate their ancestry are not interested in a mitochondrial Eve of 160,000 years ago, or a Y Adam of 60,000 years ago. They want to find cousins, living now or recently. Fortunately, dozens of websites enable genetic genealogists to take advantage of the same techniques that researchers use—tracking shared genetic markers in the form of DNA sequence variants or numbers of short repeated sequences. The reasoning in finding great-aunt Sally is the same as tracking the early expansions from Africa—the more markers (DNA sequences) two people share, the more recently we infer that they shared an ancestor. The rarer the markers, the stronger the match, just like the two students at the dorm wearing the same obscure tee-shirt.

Fast forward to March 2009. It is the 5th International Conference on Genetic Genealogy in Houston, Texas, held by FamilyTreeDNA, one of the first genetic ancestry testing companies. Knots of chattering attendees festoon the lobby of the hotel, all sporting nametags, some wearing bizarre hats or kilts. These are group administrators, in charge of organizing people linked by surnames and/or DNA sequences. FamilyTreeDNA, the brainchild of Bennett Greenspan, was founded in 2000 and by now has amassed some 220,000 results of DNA ancestry tests, including data from the Genographic Project.

Genetic ancestry companies sell tests for mtDNA and Y chromosome DNA sequences, called **haplogroups**, that serve as markers, or signposts, in the genome. Haplogroups are listed and annotated in public databases such as GenBank and published in the scientific literature. The companies sequence the markers and send the customer maps with color-coded references to major parts of the world where one's ancestors might have lived. Most people who order the tests are already in contact with others who share their surnames, memories, or information in documents such as birth, death, marriage, and baptismal certificates. For example, two people with other similarities who share 65 of 67 Y chromosome markers are extremely likely to be close relatives. If they share only 54 of the 67 markers, then they shared an ancestor farther back, the less-than-perfect match reflecting input from other genomes over the generations. FamilyTree clients sort themselves into "DNA projects" based on their haplogroups. So far, more than 84,000 surnames in the database have birthed 5,000+ projects that focus on particular DNA-defined lineages or distinct parts of the world. The genetic genealogy groups are an eclectic bunch, representing countries, surnames, and social groups that include the Mayflower Society, the African American diaspora, and several Jewish groups.

The group administrator, or leader, of the Cloud DNA Project, for example, is, like many genetic genealogists, a master of both standard sleuthing and the science behind DNA testing. He speaks excitedly, if anonymously:

> Starting a DNA project for my surname has provided several useful results. We now know that there are several distinct family lines with that surname, and the DNA tests tell us who belongs to which line. It has proven that my own line, that of Jeremiah Cloud, who was born about 1784 in Twiggs county, Georgia, is descended from the line of William Cloude who was born in 1621 in England. A Quaker, he came to America in 1682 with William Penn.

DNA information can bring surprises, as he says:

> It has been disappointing when DNA shows people belong to a different family line than they thought. It has disproven one family's oral history that they were Clouds, and has brought a person into our family who had an entirely different surname.

A practical aspect of DNA testing, he adds, is narrowing down relevant branches on one's family tree, saving time and money spent researching lines to which one is not related. The DNA approach has been especially helpful in finding genetic roots for adopted individuals.

No matter how compelling DNA results are, they do not substitute for tried and true sources of genealogical information. Nor are they infallible. For example, the same mutation can arise in unrelated families, just by chance, leading to wrong conclusions unless surnames, documents, or oral histories back up the link. Testing several parts of the genome minimizes this risk by providing more points of comparison. Another reason that DNA evidence can be disappointing is that it identifies lineages, and not individuals, and often cannot pinpoint the exact generation in which a shared ancestor lived. Also, setting DNA clocks is not an exact science. Says the Cloud family administrator,

> When we started our project, the calculations caused us a lot of grief— they showed that my third cousins and I were not related within thousands of years. We all had excellent documentation and knew we were related, so we were baffled. As other people joined the project, we were able to prove that our lines have experienced higher than average mutation rates.

Hence, the differences.

Some of the attendees at the FamilyTreeDNA conference readily admit to having been intimidated when they began considering delving into DNA science, but, after attending a few meetings, could easily chat about Y and mitochondrial DNA haplogroups. In fact, embracing DNA information can become less daunting than sifting through old photographs and records. For example, swabbing a bit of DNA from one's mouth is easier than completing the eight-page application to join the Royal Bastards, a group technically termed the "Descendants of the illegitimate sons and daughters of the Kings of Britain." Joining the group requires documentation of every single fact in the family lore. Alas, putative Royal Bastards willing to submit DNA to confirm their suspected lineage recently ran into two difficulties, one biological, one technological. First, most lines of bastards had "daughtered out," stopping with a female that halted passage of the regal Y. The second reason was technological: websites to which Royal Bastards attempted to submit information rejected entries reeking of profanity, such as the term "bastard."

Genetic genealogy can solve mysteries of history. Several of the FamilyTreeDNA groups are seeking evidence to support their descent from the so-called "Lost Colony," a small group of English settlers who came to Roanoke Island in North Carolina from 1585 through 1587. If they survived, they likely mixed with native peoples. When John Smith arrived in Virginia in 1607, natives talked about people with light skin who spoke English living close to the settlement. Coins, rings, and muskets found in the region date to sixteenth-century England. Roberta Estes, a genetic genealogist and group administrator for several DNA projects at FamilyTreeDNA, has records of Lost Colony sightings, including one from a man reporting to Smith of "certain men clothed like me" and another who referred to "an abundance of brass and houses walled like ours." Smith reported having met a ten-year-old blond "savage."

Another intriguing group united by genetic genealogy traces its roots to the 1600s in Nova Scotia, where French settlers had "AcadiaAmerIndian" offspring with native AmerIndian women. In 1755, the British chased these people from Canada, and they went to Louisiana. They are the Cajuns. Mitochondrial DNA sequences have reconnected modern Cajuns to their maternal Native American ancestors.

DNA ancestry testing unites people. A population group tragically torn from its genetic roots is formed of Holocaust survivors. The DNA Shoah project, headed by Michael Hammer at the University of Arizona in Tucson, is collecting DNA on cheek swabs from Holocaust survivors and their descendants. The growing database is being used to reunite families and to link orphans to survivors, and it may eventually be used to identify remains at the concentration camps and in museums.

TODAY: GETTING TO KNOW OUR GENETIC SELVES

This advert has run in news, entertainment, and women's magazines since spring 2009:

> Genifique: Youth Activating Concentrate. At the very origin of your skin's youth: your genes. Genes produce specific proteins. With age, their presence diminishes. Now, boost genes' activity and stimulate the production of youth proteins.

With that single, shiny, widely read page, the field of human genetics catapulted from rare appearances in the clinic and classroom to everyday experience—a skin care ad. It was accurate: the levels of proteins in our cells change whenever we do something, including aging. Not only the gene variants we inherit, but also the fluctuations in their activities as we go about living, provide a new way of viewing our physiology.

The field of human genetics has evolved from a rather obscure, largely academic discipline in the second half of the twentieth century to the subject of blogs and talk shows, and now even advertisements. In the classroom, genetics once focused on the pea plant experiments that demonstrated the pattern of trait transmission from generation to generation, discussed in Chapter 2. Doctors sought genetic explanations for the "zebras" they encountered, collections of symptoms so unfamiliar that the usual "horses"—common diagnoses—didn't seem to fit.

Genetics entered the mainstream with two technological feats: the human genome project and the HapMap project (Venter et al. 2001: 1304; International Human Genome Sequencing Consortium 2001: 860). The genome project revealed the sequence of the 3.2 billion DNA building blocks that specifies a human; the HapMap project identified places in the genome where we vary. If the genome project is compared to Google Earth maps of every street in a country, then the HapMap project just shows a few areas in each region, enough points of comparison to get a flavor of the variety of villages, towns, and neighborhoods.

Slowly, genetics is becoming the new foundation of medicine. Even diseases that are not directly inherited involve the expression of genes, such as those that guide the inflammatory and immune responses to injury and illness. A generation ago, human genetics was condensed into a few weeks of a single course in a typical medical school curriculum; today, new medical students are already familiar with DNA, RNA, and protein, and molecular genetics increasingly explains pathology. Still, the average person's visit to the average doctor is unlikely to yield much discussion of genetics, other than filling out a family history. But genetics is becoming an important part of health care, because patients are often ahead of their doctors, thanks to the exploding new market of direct-to-consumer genetic tests. Dozens of companies that offer them, over the Internet, have sprung up over the past few years, like mushrooms dotting a lawn after a soaking rain.

Although researchers are still sifting through the human genome sequence to identify genes and discover how they affect health, a robust DNA testing industry has emerged, presenting itself as an information service. A person need only select a few DNA-based tests from a website, order a kit, swish some mouthwash and spit in a tube or rub a probe on the inside of the cheek and pop it in a tube, mail the sample, and, a few weeks later, receive an analysis. Because the offered tests are not promoted as medical, they are not (yet) subject to government regulations in many places.

Traits tested range from serious diseases to characteristics that a glance in a mirror can reveal, such as eye color or pattern baldness (Table 1.4). The reliability of the tests also varies greatly, from time-tested offerings for well-studied, single-gene disorders, to extrapolations from population-derived "associations" that are so new that they could fall apart with the addition of new data by the time the person receives the results. Some geneticists fear that, without regulation, consumers could make life-altering decisions based on not-quite-solid information. The companies offering the tests counter that the information can be used to live a healthier lifestyle.

Although the genetic testing companies, at the time of this writing, do not consider test results to be equivalent to diagnoses, some consumers still fear the misuse of their genetic information. In the United States, the Genetic Information Nondiscrimination Act (GINA) was passed in 2008 and offers some protection, prohibiting employers and health insurers from requiring genetic testing. The actual effect of all of this testing on preventing disease or maintaining health has yet to be evaluated.

Table 1.4 Some direct-to-consumer genetic tests

Age-related macular degeneration	Heart attack
Alcohol flush reaction	HIV resistance
Alzheimer's disease	Lactose intolerance
Asthma	Lupus
Athletic performance	Malaria resistance
Bitter taste	Multiple sclerosis
Brain aneurysm	Muscle performance
Cancers	Nicotine dependence
Crohn's disease	Norovirus resistance
Deep vein thrombosis	Obesity
Diabetes	Osteoarthritis
Earwax type (dry or wet)	Parkinson's disease
Essential tremor	Psoriasis
Eye color	Restless legs syndrome
Gallstones	Rheumatoid arthritis
Glaucoma	Ulcerative colitis
Gout	

OUR FUTURE

In the days before downloads, noted writer and paleontologist Stephen Jay Gould once compared evolution with a tape and asked, what if we could run the tape backwards and then start again, making a different set of choices? Would humans still come to exist, or would the planet be overrun with monkeys, as it was 12 million years ago, before the first inklings of humanity appeared? Or might australopithicines still rule? If another type of animal like us had evolved, alongside us but distinct enough not to mate with us, how would it differ from us?

Likewise, we can ask, where might evolution take us?

A difference between running the tape of life forward from now, as opposed to running it forward from the beginning, is the unprecedented degree of control that our species has over the environment. The seeds of molding the environment grew from our ancestors' abilities to think, plan, and cooperate. *Homo* eventually came to dominate the australopithecines, we think, because of more sophisticated tools, more enduring family structure, and the use of fire. When agriculture began to sweep the globe, the enclaves of indigenous peoples began to shrink, taking with them many valuable clues to the past.

Being in the right or wrong place at the right or wrong time has also played a role in destiny. Throughout prehistory, bands of peoples came and went, their fates sealed by what they could do with their genomes, but also affected by happenstance.

In the distant past, nature was the chief agent of selection. Those fortunate enough to live in a time or place where their inherited characteristics were a help and not a hindrance prevailed, passing on those very gene variants that enabled them to survive. Today, we provide or control many of those selective forces, in the form of opportunities and access to services. A newborn would likely face different heath challenges in her lifetime, and die of a different cause, if born today in Malibu, Mumbai, Mozambique, or Madagascar. Her fate would depend as much on her genes as it would on her wealth, education, and physical environment.

Envisioning the future human gene pool has long piqued our imaginations (Table 1.5). Novels and films explore "dystopic" societies that choose who can reproduce; circumstances in which

Table 1.5 Future views of humanity

Time	Novel/film	Plot
soon	*GATTACA*	An "invalid" man, whom society keeps from fulfilling his dreams because he is destined to die young and is myopic, buys DNA from a "valid" individual, only to be exposed by the DNA in an errant eyelash.
2274	*Logan's Run*	In a domed city following a global holocaust, a computer in control allows people to live until age 30. Sex is free, everyone is young and beautiful, but the protagonists escape and find an old man.
2540	*Brave New World*	In the World State, everyone lives a happily drugged existence, with no individuality or families. Five castes are gestated, their intelligence controlled by chemicals, not genes, in Hatcheries and Conditioning Centres.
3978	*Planet of the Apes*	Over centuries of people using apes for menial tasks and medical experimentation, people get lazy, and the apes smart. After DNA damage and intense selection by a nuclear war, humans go wild and become slaves of their hairy cousins.
802,701	*The Time Machine*	A Time Traveller visits a time after humanity has split. The above-ground Eloi are blonde, thin, meek, placid, and unthinking. The below-ground Morlocks are dark, stocky, aggressive, and use technology to cultivate the Eloi as food.

humans have diverged into separate but very unequal species or come under the domination of apes; and starting our species anew from a select few following a fill-in-the-blank scenario of meteor strike, alien attack, or exploding sun. A book and TV series have even explored a world without people. These fictional works look ahead at the possible implications of the most controversial areas of the field of human genetics today—controlled reproduction, eugenics, fear of technology, personal identity, and, perhaps most important, the pervasive idea of **genetic determinism**, that we are our genes. Which, if any, of these scenarios might come true? Several, for we can and do change our genes, mostly at the population level.

Historically, we have objected to intentional attempts to alter future gene pools, called **eugenics**, yet we make similar decisions at the individual level. In the film *GATTACA*, future parents select the traits of their offspring-to-be; today, people use a technology called preimplantation genetic diagnosis to have a child whose tissue and organs can save another (see the box entitled "What would you do?"). Genocide—the mass killing of certain types of people—still alters gene pools. Widespread use of technology to skew the sex ratio in China, begun in 1970, has led to an unbalanced society now, where the tables have turned. Today, the scarcity of daughters, sisters, wives, and aunts has increased their perceived value. Successful gene-based treatments may perpetuate gene variants that, if nature took its course, would gradually be weeded out of populations.

Like the fortunate few selected to survive in the fictional dystopias of the future, the ways that we live will affect the course of our own evolution. What imprint will disparities in access to health care leave in the future human gene pool? Will stem cell and gene therapies only be available to people who happen to live in the nations where the research was encouraged early on, or who live elsewhere and can afford the best care? If we learn to prevent or treat today's big killers, cancer and heart disease, which other conditions will then kill most of us? At the same time, will accumulation of environmental toxins hike the mutation rate, introducing genetic disease anew? Or will natural selection present new challenges that we cannot even imagine?

What will we look like thousands of years hence? Will the ease with which we travel and meet people on the Internet eventually mix us so much that we all come to have a uniform, global human gene pool, obliterating the distinctive collections of gene variants that once defined the displaced Cajuns, Ashkenazi Jews, and Native Americans, and the world's remaining indigenous peoples? Will the boundaries and separations that defined human populations in times past gradually vanish as we assimilate into one mass of humanity? We don't know.

Human genetics is perhaps unique among the life sciences in that it gets to the fundamental question of just who and what we are— where we came from, and perhaps what we can become. It is all written in that personalized instruction manual, the human genome.

WHAT WOULD YOU DO?

Late in the summer of 1999, Adam Nash was born in Chicago. Nine months earlier, researchers had mixed in a lab dish the eggs and sperm that would produce Adam. A few embryos grew in the dish, until each consisted of eight cells. Then, the researchers plucked off one cell from each of the embryos and analyzed the DNA. One sampled cell came from an embryo that would develop into a child who would be a tissue-match for six-year-old Molly Nash. More importantly, that child would not have the disease that was making Molly so sick, Fanconi anemia. So the researchers took the seven-celled counterpart to the sampled cell and implanted it gently into Molly's mother's uterus. It became Adam.

A month after Adam's birth, his umbilical cord stem cells were infused into Molly, saving her life. The parents, Lisa and Jack, made the rounds of talk shows. At first they faced great criticism—much as the family of Louise Joy Brown had, who in 1978 became the first "test tube baby." Even as the public became more accustomed to the technique that saved Molly Nash, called preimplantation genetic diagnosis (PGD), controversy remained. In 2004, Jodi Picoult wrote *My Sister's Keeper*, a novel in which the protagonist was, like Adam Nash, conceived to provide tissue for her ailing sister—only she objects to continually donating her parts.

Despite the continuing discussion, thousands of children have been spared their family's inherited disease thanks to PGD. The technique has become so popular that it is now routinely used to "screen" embryos fertilized in the lab (*in vitro* fertilization (IVF)), even if there is no family history of inherited disease, to select out embryos with abnormal chromosomes. But PGD and IVF could do much more, as we now know the sequence of the entire human genome.

Where will the use of PGD stop? Are there any limits? Bioethicists speak of a "slippery slope," a situation without guidelines that veers out of control. This may happen with the wedding of two technologies— PGD and the ability to detect many, if not all, of our gene variants.

What limits would you place on the specific traits screened for and used to select your children? What facts would you want to have before you decided to use PGD to select a future child who inherits gene variants associated with high intelligence, longevity, or athletic prowess?

SUGGESTED READING

Many books, articles, and websites explore humanity's past. Two articles that specifically chronicle the stories in the sediments of the Awash River region of Ethiopia are "Human origins from Afar," by Ricki Lewis, published in *The Scientist* in the April 12, 2004 issue (18(7), cover story), and "Rocking the cradle of humanity," by Elizabeth Pennisi, in *Science*, 319: 1182–1183. The October 2, 2009 issue of *Science* magazine has several articles on *Ardipithecus*. *Lucy: The Beginnings of Humankind*, by D. Johanson and M. Edey (Simon and Schuster, 1980) describes the discovery of the australopithecine Lucy. The Talk Origins website (www.talkorigins.org) offers lively discussion of non-scientific ideas about the origin of humanity.

For more recent ancestors, consult the Genographic Project website (https://genographic.nationalgeographic.com/genographic/index.html); *The Journey of Man: A Genetic Odyssey* and *Deep Ancestry: Inside the Genographic Project*, by Spencer Wells (Random House, 2004 and 2007); and *The Seven Daughters of Eve*, by Bryan Sykes (W.W. Norton, 2002). The International Society of Genetic Genealogy (www.isogg.org/) provides information on ancestry testing. *The World Without Us*, by Alan Weisman (St. Martin's Press, 2008), is a look ahead to a time after humans have become extinct.

2

FROM MENDEL'S PEAS TO DOUBLE HELICES

Modern genetics began in 1900, with the discovery of Gregor Mendel's paper reporting the two basic laws of inheritance, but people were aware of the passage of traits from generation to generation long before Mendel carried out his famous experiments on pea plants.

THE IDEA OF HEREDITY

The earliest thoughts about heredity required assembling three types of observation:

- family structure;
- natural variation in traits;
- trait variations repeating among members of a family.

People may have put these pieces together long before the science of genetics was born, perhaps at many times and in many places.

FAMILIES

About 3.7 million years ago, a family walked in what is today an archeological site in Tanzania called Laetoli, in the shadow of a volcano 20 kilometers away. In 1976, two researchers in the same area were tossing elephant dung when they stumbled upon 70 human-like footprints extending about 9 meters, preserved in volcanic ash. The footprints, some within others, came from two individuals, one larger and one smaller, walking abreast, with smaller impressions from a third individual behind. The prints from the smaller of the two adults listed to one side—were they from a mother carrying an infant on her hip?

As with most fossils, unusual environmental conditions captured the moment: a light, feathery rain gently coated the impressions in the ash. Over time, the prints hardened into rock called tuff. The location, the dating of the footprints, the walking speed deduced from the print spacing, and the absence of knuckleprints suggest that the small group were *Australopithecus afarensis*, like little Lucy described in Chapter 1. They lived in family units—requirement one for recognizing human heredity.

NATURAL VARIATION

Prints and bones do not reveal much about human traits, but DNA does. For example, DNA evidence challenges the long-held view of Neandertals as dark-eyed and dark-haired, indicating that some of them had red hair, fair complexions, and possibly freckles. This new view came from a leg bone discovered by graduate student Araceli Soto Florez, from the University of Oviedo, on a dig in 2000 in a "tunnel of bones" in El Sidron, in northern Spain. She carefully collected the precious bone fragments and placed them in a bag that was transported on ice to a laboratory at the Institute of Evolutionary Biology in Barcelona. The El Sidron DNA revealed a mutation in a gene called *MC1R* that provided red tones to the hair and skin (Laleuza-Fox et al. 2007: 1453). A family of dark Neandertals would have certainly noticed a sunny-colored newborn. Other evidence suggests that the freckled redheads appeared near the end of the Neandertals' existence. From 38,000 to 70,000 years ago, the Neandertal population fell, with perhaps fewer than 3,500 breeding females. Analysis of six fossil Neandertals found far apart in Europe showed that their genomes were so similar that inbreeding was nearly certain, increasing the incidence of the recessive redness.

We can only wonder how the Neandertals reacted to those who looked different. We do know that, much later, the ancient Egyptians revered people with another inherited variation—achondroplasia, or short-limbed dwarfism. It is passed on in an **autosomal dominant** manner. This means that males and females are affected (autosomal) and that each child of an affected parent has a 1 in 2 chance of being affected (dominant). From 4500 to 3000 BC, Egyptians with achondroplasia held a variety of respected jobs, and depictions in paintings and on statues, vases, and tombs celebrate their daily activities. Preserved skeletons of people with achondroplasia in Egypt from 2700 to 2190 BC have been found in fancy burials sites. Dwarfs even achieved the status of the divine in this culture, in the forms of the gods Bes (protector of women and children) and Ptah (regeneration). Again, the environment in ancient Egypt—very dry air and both natural and aided mummification—preserved these vignettes.

FAMILY RESEMBLANCES

Evidence of heredity was everywhere, long before the Egyptians paid tribute to their little people. Agriculture provided the numbers that revealed the laws of inheritance, because the controlled breeding that is the backbone of farming is actually a genetic experiment.

From 15,000 to 8,000 years ago, agriculture arose independently in different parts of the world. The time of its origin depended upon such factors as climate, geography, the movements of hunter-gatherers, and their ultimate ability to control the environment. Agriculture grew out of astute observations. Imagine a group of nomads settled overnight in an area with edible plants. They ate, and then perhaps took food with them when they left, leaving seeds behind from tasty fruits and grains. A year later, the people returned and noted the same types of plant growing. At some point, someone observed a seedling sprouting a tiny stem and linked the seeds of one season to the food of the next. Then, people began leaving seeds on purpose, saving the seeds from the most robust plants each season to begin the crops of the next. Domestication of wild animals such as sheep and goats paralleled the control over plant breeding. Eventually, people must have realized that, if they could grow crops, they needn't move to find food. Settlements began.

The first farmers were insightful observers of natural variation. Consider the origin of corn, also known as maize and *Zea mays*. In the Balsas River basin in southern Mexico, about 9,000 years ago, people began, unknowingly at first, fashioning the seeds of a wild grass that still grows there, called teosinte, into corn.

Corn and its ancestor, teosinte, do not look very much alike. In contrast to a corn plant's majestic height, with its seeds tucked into paired protective husks and its stalk firm enough to withstand a strong wind, teosinte is stout and branched, each outgrowth holding a small, ragged ear. Corn kernels are stuffed with protein, oil, and starch, in a soft covering that teeth can easily tear. In contrast, teosinte kernels are small, trapped in a hard fruitcase, and form two scraggly rows in a poor excuse for a cob. Whereas the corn's husk hugs the succulent kernels in their neat rows, the stone-like teosinte seeds simply drop and scatter. Teosinte's tough seeds, though, reflect natural selection—they traverse an animal's digestive tract unscathed, plopping from the anus into hopefully fertile ground.

Corn couldn't have evolved naturally, for its seeds would never survive the journey through the digestive tracts of herbivores. So how has corn persisted? We helped it.

Archeological evidence indicates that people once grated teosinte seeds into flour. In doing so, someone must have come upon an unusual plant that had larger or softer kernels. Perhaps from knowing how to cultivate other crops from early agriculture in Mexico—yams, squash, cotton, peanuts, and peppers—the people knew to save the seed and plant it to pass on the desirable traits. Over time, with continual selection at each generation for the most palatable and digestible kernels, teosinte became corn. However, in the process, it sacrificed its autonomy, because people would always be required to remove the kernels from the husks and plant them.

Selecting seeds to perpetuate a valued trait is recognition of genetics. Today, we know that the differences between corn and teosinte come down to a mere handful of genes. A gene called *tb1*, for example, controls body form—the variant in teosinte confers the shrub-like shape, whereas the variant in corn suppresses an ancestral tendency to grow laterally, resulting in the cornstalk. Another gene controls the pattern of deposition of silica (the main component of sand) and lignin (a carbohydrate) in the kernels, which is responsible for the differences in hardness between the seeds of corn and teosinte.

Comparing gene sequence differences among modern cultivars of corn and teosinte and considering mutation rates and geographical distribution suggest a single origin for corn about 9,000 years ago, in southern Mexico. Consistent with the DNA story are preserved ears that track the changes that accompanied corn's human-aided evolution from teosinte. Once corn had been cultivated, it rapidly replaced its forebear. Evidence of tools to process corn date from 8,700 years ago—tiny crevices in the tools bore starch characteristic of modern corn, but not of teosinte.

Agriculture began in different parts of the world within a few thousand years and transformed humanity. The hunter-gatherer lifestyle gave way to settlements, and people came to rely on fewer varieties of foods as crops abounded. So, too, were the seeds sowed for the birth of the field of genetics.

MENDEL'S EXPERIMENTS—THE MYSTERY OF THE MISSING TRAITS

The two basic laws of inheritance emerged from people puzzling over a common phenomenon in their gardens and farms: vanishing and reappearing traits. Specifically, crossing plants where each has a different version of a trait—flower color, for example—produces hybrids with flowers of only one of the colors. However, cross those hybrids, or allow them to self-fertilize, and, in the next generation, both flower colors appear. Why? And how does a trait disappear in the hybrid generation? A monk named Gregor Mendel (1822–1884) used math to solve the mystery of the missing traits.

MENDEL THE MAN

Johann Mendel was born in 1822 to serf parents, in a tiny village in northern Moravia. Johann would become "Gregor" in 1843, and Moravia would become Czechoslovakia in 1945. Mendel was a good student, and so the schoolmaster asked him, at age 12, to take the entry exam for the Imperial Royal Gymnasium. After success at this "grammar" school, he briefly attended university, and then, in 1843, entered the monastery of the Augustinian Friars at Saint Thomas in Brunn (now Brno) as a novice. There he studied theology, history, Greek, law, and teaching. After being ordained in 1847, Mendel

32

math and Greek at a local school, but failed his teacher
..tion exam because he couldn't classify mammals. For a short
time, as a parish priest, he visited patients in hospitals, but became
so squeamish, in those days before anesthesia, that he ruled out a
career in medicine.

Despite his ineptitude at classifying mammals, Mendel proved
himself a gifted teacher, and the monks, recognizing this, sent him
to the University of Vienna. There he studied botany, physics, and
math. Still without a degree, Mendel returned to Brunn in 1853 to
teach natural history and physics, which he would do for 14 years.
Shortly after his arrival, he again took the teaching certification
exam, which he failed due to a botany question. Ironically, this was
also when he started breeding plants in back of the monastery.

EXPERIMENTS REVEAL THE LAWS OF INHERITANCE

To investigate how traits vanish in the hybrids, Mendel chose seven
"differentiating characters," each with two obvious expressions, in
the garden pea. He was methodical, persistent, and perhaps a little
obsessed. Mendel transferred pollen himself, so that he knew which
plants transmitted which traits. He would kneel before the chosen
plants, cut off the stalks that house the pollen, then collect and paint
the pollen onto the recipient female plants using a small artist's brush.
He'd place a paper bag over the altered plant and scribble in a tattered
notebook the trait variants he had crossed.

In the first set of experiments, Mendel crossed plants that
exhibited alternate expressions of a trait, such as round or wrinkled
seeds, tall or short stems, or yellow or green peas. Much of the
language of Mendelian (single-gene) inheritance came later, but
introducing it provides a way to follow the experiments. A particular
variant of a gene is called an **allele**. The expression of an inherited
trait is called its **phenotype**—what it looks like. The specific alleles
in an individual constitute the **genotype**.

Mendel tested seven distinct traits in all. The initial crosses
established which allele was **dominant** and which **recessive** for each
pair. He did this by breeding tall plants for many generations, until
they only yielded tall plants, and he did the same for short plants.
Then, he'd cross a pure, or "true-breeding" tall plant to a true-
breeding short plant to obtain the hybrids. The dominant expression

of a trait is the one that appears in the hybrid generation, whereas the recessive expression disappears in the hybrid generation, but reappears in the generation after the hybrids. Traits reappear, Mendel reasoned, because particles transmit them, and the particles persist. For traits with only two alleles, the dominant is designed with a capital letter and the recessive with a small letter, both italicized.

Mendel next self-fertilized hybrids. This experiment is called a **monohybrid cross**, because it follows one trait. When Mendel catalogued the offspring of the self-crossed monohybrids, a pattern emerged: the dominant trait outnumbered the recessive trait three to one. For example, in one set of experiments, tall hybrid plants, self-crossed, yielded three tall offspring to every one short offspring (Figure 2.1 and Table 2.1). The fact that the sets of experiments for each of the seven traits yielded a similar ratio suggested a common mechanism.

In Mendel's first set of experiments, he crossed pea plants that "bred true" for two expressions of the same trait, then self-crossed the resulting hybrids. He interpreted the 3:1 ratio in the second generation to represent the segregation of alleles for genes transmitted on different chromosomes.

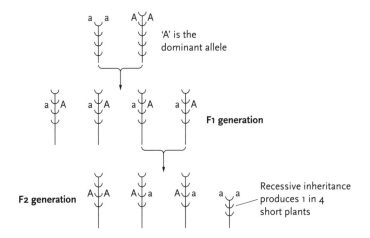

Figure 2.1 A monohybrid cross

Source: Adapted from *Genetics for Healthcare Professionals*, Figure 19, p.59

Table 2.1 Derivation of the 3:1 phenotypic ratio of a monohybrid cross

Parents	Sex cells	Offspring combinations	Phenotype
Aa	A, a	AA (1 way)	Tall
		Aa (2 ways)	Tall
		aa (1 way)	Short

Phenotypic ratio: 3 tall:1 short

Genotypic ratio: 1 AA:2 Aa:1 aa

If A is the allele for tall plant height and a is the allele for short plant height, then a monohybrid cross yields 3 tall plants for every 1 short plant.
A genotype of Aa arises in 2 ways because either parent can provide the dominant (A) or the recessive (a) allele.

Mendel continued these first experiments a generation further, which confirmed that the dominant allele masks the recessive allele. Crossing short offspring of the tall hybrids to each other produced only short plants. But, crossing the tall plants in the generation after the hybrids to the rarer short plants revealed that the tall plants themselves came in two varieties: one-third of them "bred true," producing only more tall plants, but two-thirds produced tall to short offspring in the ratio 3:1, as if they themselves were hybrids. They were. This first of Mendel's generalizations, which would be named the "law of segregation," states that inherited characters are present in two copies, which separate as sperm and egg form, and then come together in predictable combinations—assuming the two alleles are present in equal numbers and join at random.

New terminology eventually replaced Mendel's "hybrid" and "true-breeding" (Table 2.2). The hybrids are called **heterozygotes** and have two different alleles of a particular gene (Figure 2.2). The true-breeding individuals that transmit the same phenotype (expression of a trait) generation after generation are called **homozygotes**. They have two identical alleles of a particular gene. Although one individual—pea or person—has only two alleles per gene, which can be the same or different, most genes have more than one allele, so that several types of heterozygote are possible. Chapter 3 describes this situation vividly for the disease cystic fibrosis, which varies greatly in severity.

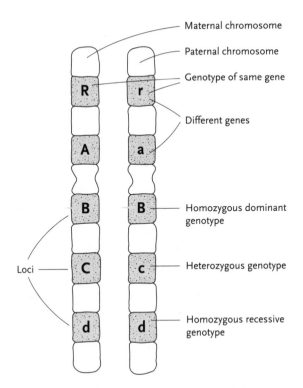

Figure 2.2 The relationship between genes and chromosomes. Each chromosome carries hundreds of genes

Source: Adapted from *Introducing Genetics: From Mendel to Molecule*, Figure 2.2, p. 11

Table 2.2 Mendelian terminology

Designation	Genotype	Phenotype	Example
heterozygote	Aa	dominant	round peas
homozygous recessive	aa	recessive	wrinkled peas
homozygous dominant	AA	dominant	round peas

In a second set of experiments, Mendel followed the inheritance of two and then three traits at a time. He crossed plants that make only round, yellow peas with plants that make only wrinkled, green peas. The pods of the hybrid generation had only round, yellow peas. Round must be dominant to wrinkled, and yellow dominant to green, he reasoned. Mendel assigned letters to the traits: *A* was round and *a* wrinkled; *B* was yellow and *b* green. The parents, therefore, were *AABB* and *aabb*, and the first-generation plants—called **dihybrids** because they were hybrid for two genes—were *AaBb*.

Self-fertilization of the dihybrids (*AaBb*) led to plants with pods containing four varieties of pea. Most were round and yellow like the dihybrids (*A_B_*), but a few were the other three possible combinations: round and green (*A_bb*), wrinkled and yellow (*aaB_*), and wrinkled and green (*aabb*) (Table 2.3). After observing many crosses, Mendel deduced that the four classes of offspring from a dihybrid cross appeared in an approximate 9:3:3:1 ratio. What would become known as Mendel's "law of independent assortment" explains it: "the relation of each pair of different characters in hybrid union

Table 2.3 Derivation of the 9:3:3:1 ratio of a dihybrid cross

Parents	Sex cells	Offspring combinations
AaBb	*AB, Ab, aB, ab*	*AABB* (1 way)
		AaBB (2 ways)
		AABb (2 ways)
		AaBb (4 ways)
		9 ways
		AAbb (1 way)
		Aabb (2 ways)
		3 ways
		aaBB (1 way)
		aaBb (2 ways)
		3 ways
		aabb (1 way)

Genotypic ratio: 9 A_B_:3 A_bb:3 aaB_:1 aabb
Phenotypic ratio: 9 round, yellow:3 round, green:3 wrinkled, yellow:1 wrinkled, green
(The different ways reflect the fact that a particular allele can come from either parent)

is independent of the other differences in the two original parental stocks" (Mendel 1865). Critical to the finding was that the traits Mendel studied represented genes carried on different chromosomes. The results would have been different had the genes been "linked" on the same chromosome.

Over eight years, Mendel analyzed 16,384 plants. He concluded, "hybrids form egg and pollen cells of different kinds, and that herein lies the reason of the variability of the offspring." His genius was that the chromosomes that separate the "characters," as sperm and egg form in meiosis, had not yet been described.

OVERLOOKED AND THEN DISCOVERED

Mendel presented his work in talks in February and March of 1865 and published his paper in the proceedings of the Brunn Society for the Study of Natural History (Mendel 1865: 3). A flattering local newspaper article was the only immediate reaction. Disappointed, he left his beloved garden in 1868 to focus on monastery administration. However, like many people sidelined before they planned to retire, Mendel found plenty to do. He led the local volunteer firefighters and became fascinated with the weather, even writing a report about a tornado that ripped up the monastery.

Mendel's paper went to 120 libraries, including some in the United States and the United Kingdom. A few papers referenced, but misinterpreted, his findings. Fortunately, though, reprints of his paper passed, silently, among a few scientists. It remains a mystery why Mendel's elegant experimental work went unnoticed at the same time that Charles Darwin's much longer and mostly obser-vational *Origin of the species* became an instant sensation. Perhaps it was math anxiety. H.G. Wells, Julian S. Huxley, and G.P. Wells wrote, in their 1929 textbook *The Science of Life*, that Mendel's presentation "dealt chiefly with peas and arithmetic, not the sort of things that cause excitement and clamour, and in the confused tumult of the nineteenth century evolution controversy, they passed unnoticed."

His genius unknown, Mendel died, according to Wells, Huxley, and Wells, in 1884, a perplexed and sad man. But the sparse citations and passing around of the paper would eventually, at the turn of the century, come to the attention of three botanists who had independently discovered Mendel's laws. Preparing to publish their own work, Hugo De Vries (1848–1935), Carl Correns (1864–1933), and Erich von Tschermak (1871–1962) scouted the literature and came upon Mendel's paper. English embryologist William Bateson (1861–1926) translated Mendel's paper into English in 1901, drawing many more readers.

With the recognition of Mendel's contributions, a new lexicon to describe heredity began to emerge. Danish botanist and geneticist Wilhelm Johannsen (1857–1927) contributed *genotype* (the gene variants) and *phenotype* (their observable expression) in 1903, when he called the still-intangible units of inheritance "theoretical hereditary corpuscles." In 1909, he coined the phrase "genes," and Bateson contributed "genetics."

Experimental data confirming Mendel's laws began to accumulate in mice, guinea pigs, chickens, rabbits, and fruit flies, among others. Natural variation of all types was now seen to reveal ratios: from dwarfism to flower color to cattle horns and attached earlobes. Mendel's laws are laws because they apply to all organisms. At the same time, cytologists—the early cell biologists—were closing in on the chromosomes. Yet a third thread, the chemical nature of the gene, would soon entwine with the other two to weave the fabric of the fledgling field of genetics.

THE CHROMOSOME CONNECTION

A dozen years after Mendel published his paper, researchers elsewhere were discovering the chromosomes that would provide the physical explanation for the transmission of traits and genes. Chromosomes are the dark-staining structures in cell nuclei that appear as rods when the cell is dividing, but at other times are so unwound that they cannot be seen. Both chromosomes and genes come in pairs and separate at each generation, then unite in random combinations. Genes are part of chromosomes—but that was not initially obvious.

GREAT MINDS THINK ALIKE (PART 1)

The earliest known sketches of chromosomes came from German biologist Walter Flemming (1843–1915). In 1878, he observed tangled chromosomes in cells from salamander gills and fins and named the mysterious substance *chromatin*. He did not know of Mendel's work of 13 years earlier. German anatomist Wilhelm von Waldeyer-Hartz (1836–1921) originated the term *chromosome* in 1888.

By the turn of the nineteenth century, the colored threads of chromatin were known to divide lengthwise during both types of cell division (mitosis and meiosis) and pass in constant number to daughter cells. **Mitosis** is the division of body (somatic) cells. **Meiosis** is the type of cell division that forms the sex cells (gametes)—eggs and sperm. Body cells have two sets of the 23 chromosome types and are said to be **diploid**. Sex cells have only one copy of each chromosome and are said to be **haploid**. When sperm and egg come together at fertilization, the diploid chromosome number is restored.

In 1902 and 1903, Walter S. Sutton (1877–1916), a graduate and medical student at Columbia College in New York City, and Theodor Boveri (1862–1915), in Germany, independently connected chromosomes to inherited traits. Both men observed the doubling and parting of chromosome pairs as sex cells formed during meiosis, as well as the maintenance of chromosome number as cells divided by mitosis. Sutton looked at cells from a species of giant grasshopper, while Boveri investigated sea urchins and a worm that infects horses. Mendel, Sutton, and Boveri had each deduced that, to maintain the correct amount of genetic information from generation to generation, each parent's contribution must be halved. Sutton also noted that the number of traits exceeded the number of chromosomes —therefore a chromosome must carry the units for several traits (a little like ten people wearing, altogether, 80 items of clothing). Evidence for Sutton's idea required a return to the Mendelian approach of setting up crosses and observing the progeny.

From 1905 to 1908, a series of papers from Bateson and fellow British geneticist R.C. Punnett (1875–1967) picked up where Mendel left off. Whereas Mendel chose, by chance, traits carried on different chromosomes, Bateson and Punnett followed two traits in the sweet pea that did not yield Mendelian ratios—because their genes were on the same chromosome. Bateson and Punnett crossed true-breeding plants with purple flowers and long pollen grains to true-breeding plants with red flowers and round pollen grains. When they crossed the dihybrids, instead of the expected 9:3:3:1 ratio, most of the offspring instead resembled either parent. The few exceptions arose, they deduced, when something disrupted the relationship of genes on a chromosome.

Bateson and Punnett didn't know that their observations of linked genes explained Sutton's idea that each chromosome carried several genes. The facts fell into place in room 613 in Schermerhorn Hall at Columbia College, in yet another experimental organism, the fruit fly *Drosophila melanogaster*.

THE FLY ROOM

The messy and stinky "Fly Room" was the domain of Thomas Hunt Morgan (1866–1945), who, early in the twentieth century, headed a new, experimental zoology program. Excited by Mendel's findings, Morgan grew impatient with the long time it took to conduct experiments using rodents. Fruit flies were an enticing substitute: they offered 30 generations a year, and all they required to happily reproduce were buckets of bananas and milk bottles.

The Fly Room had eight cluttered and crammed desks. Cockroaches skittered across the floor. But the sense of excitement and cooperation that permeated the place would become a model of how to run an academic science lab. Every available surface was festooned with the small, cardboard-stoppered glass milk bottles, each housing hundreds of flies. Insect escapees would often light on the researchers' hair, and the air bore the distinctive aroma of the overripe bananas mashed into milk bottle bottoms. Female flies would lay their eggs onto this goo, and out would hatch the first of three waves of larvae that would mature within it. The worm-like creatures with their perpetual motion jaws would eat themselves towards adulthood, finally encasing themselves in cocoons and, a few days later, emerging and unfolding as full-fledged flies. Half a dozen future Nobel laureates were among "Morgan's boys," and the Fly Room spawned an academic pedigree of the founders of modern genetics in the twentieth century.

An easy-to-raise animal with many offspring was not enough to study genetics, however. Morgan needed variants, akin to Mendel's short and tall pea plants. It took a few years of raising and carefully inspecting flies to find interesting new mutations, for it was not yet known how to help nature along and use chemicals and radiation to cause mutations.

One early spring day in 1910, Morgan discovered a white-eyed male fly alone among his red-eyed ("wild type") brethren. He mated the precious male with the odd eyes to his virgin sisters, a way to track trait transmission, much as Mendel hand-pollinated plants. All of the white-eyed male's offspring had red eyes. Was this another case of a trait hiding among the hybrids? To find out, Morgan crossed the red-eyed hybrids with each other and then scrutinized the offspring. If he didn't consider gender, then red eyes outnumbered white three to one—a neat Mendelian ratio. However, he quickly noticed that all of the female flies had red eyes, whereas the male flies were split—half white-eyed, half red-eyed. Then, Morgan discovered two other traits with the same peculiar inheritance pattern—yellow body and underdeveloped wings. Aware that males had an X and a Y sex chromosome, whereas females had two X chromosomes, Morgan proposed that the genes for the traits appearing only in males were carried on the X and were expressed in males because they lacked a second X that would mask the traits. Males could be white-eyed or red-eyed, depending upon which X chromosome they inherited from their red-eyed, hybrid mothers. Morgan had described X-linked inheritance, the specific pattern of trait transmission for genes on the X chromosome.

The "fly pickers" in room 613 continued to catalog new mutants. They set up crosses to see which pairs of traits assorted independently—carried on different chromosomes—and which traits tended to be inherited together. The trait pairs fell into four groups. Was it coincidence that the fly has four sets of chromosomes? No. Genes linked on the same chromosome are inherited together because the chromosomes transmit them together. The rare flies with mixed up, or recombined, trait combinations, compared with their parents, arose when paired chromosomes literally crossed over, swapping parts. This happens during meiosis.

An undergraduate researcher in the Fly Room, Alfred Henry Sturtevant (1891–1970), was intrigued by the idea of crossing over disrupting linkage. As a child growing up on a farm in Alabama, Sturtevant had drawn pedigrees of his father's horses. When he learned about Mendel's laws in Morgan's class, he became excited at applying them to the traits he had followed in horses. That led him to join the lab group.

Sturtevant thought about five traits linked on the fly's X chromosome. One night, he was struck with an idea so compelling that he cast his homework aside and stayed up until the wee hours, sketching. He took his mentor's vision of genes linked on a chromosome, but occasionally separated by crossing over, a step farther: what if the frequency of the recombinant class of offspring reflects relative distance of the genes along the chromosome? Genes at opposite ends of a chromosome would have a large recombinant class and represent a large physical distance, whereas genes close together on the chromosome would generate few recombinants, simply because there was less space for a crossover to occur. It was like considering stores on opposite sides of a street. There are more sites at which to cross between stores at either end of the block than there are between two stores in the middle of the block.

At age 19, Sturtevant drew the first map depicting genes on a chromosome (Sturtevant 1913: 43). Little did he know that it would become the foundation for all future genetic maps, even those used today to pepper the genome with a million or more landmarks.

HUMAN CHROMOSOMES

When some researchers were encountering Mendel's laws, others were examining chromosomes, not, at first, realizing that the two fields of study were closely linked. However, the colored tangles were so difficult to visualize under a microscope that these early cytologists (today called cell biologists) counted anywhere from 30 to 80 of them in human cells. The first close estimate came in 1923, when Theophilus Painter published sketches of what he thought were the 48 human chromosomes (Painter 1923: 247). Painter might not have been mistaken. He used a sample of testicular tissue removed from a patient at a Texas state mental hospital who had been castrated to quell his violent behavior. It is possible that the man indeed had 48 chromosomes, and had a syndrome responsible for his behavior. Painter also showed that human males and females have the same chromosome number—for a time it had been thought that males have one X but no Y, because the Y is very small and therefore difficult to see.

Visualizing chromosomes depended on improvement in technique, and this happened gradually. First, researchers learned how to use an extract from the crocus plant, called colchicine, to stop cells when they are in the midst of dividing. At this time, the chromatin is at its most condensed, appearing as rods. In 1951, a technician stumbled upon a way to separate the tangle of chromosomes. While trying to culture white blood cells, the technician bathed them in a salt solution that wasn't concentrated enough. With the interiors of the cells saltier than their surroundings, water rushed in, swelling them so much that the chromosomes spread apart, like water added to a small swimming pool moving floating toys away from each other. Two years later, researchers perfected ways to draw up cells into a pipette, splash them down into a droplet of stain, and then gently add a glass coverslip, so that the chromosomes spread at one level. These experiments eventually made it clear that the human somatic (body) cell chromosome number is 46, and other work revealed that gametes (sperm and egg) each have 23 chromosomes.

Staining chromosomes improved, along with the methods to separate them. At first, the stains were so generalized that all the chromosomes were the same color. Chromosome charts depicted them in size groups, such as a "D group" chromosome. In 1959, researchers associated the first syndrome with an abnormal chromosome number—the 47 chromosomes of a person with Down syndrome. In the 1970s, newer stains created banding patterns distinct to each chromosome type (Figure 2.3). Shortly after, researchers labeled individual chromosomes, first with radioactive chemicals and later with fluorescent chemicals, which zeroed in on specific DNA sequences.

Each human somatic cell has 23 pairs of chromosomes, including 22 pairs of autosomes (non-sex chromosomes) and 2 sex chromosomes (the X and the Y). The chromosomes in Figure 2.3 are from a male.

Human chromosomes are displayed in size-order charts called **karyotypes**. The chromosomes are numbered 1 through 22, from largest to smallest, plus the sex chromosomes, the X and the Y. Abnormal numbers of chromosomes or their parts usually affect health. Extra chromosomal material is better tolerated than missing material, and this is why Down syndrome, caused by an extra chromosome 21, is the most common chromosomal abnormality. Table 2.4 lists types of chromosome problems.

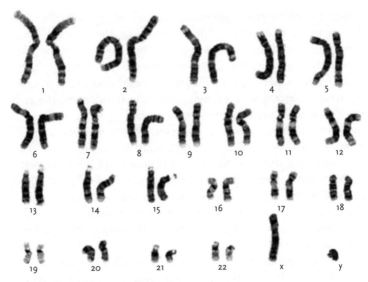

Figure 2.3 Human chromosomes

Source: *Genetics for Healthcare Professionals*, Figure 15, p. 56

Table 2.4 Types of chromosome abnormality

Description	Technical term
An extra entire set of chromosomes	Polyploidy
An extra individual chromosome	Trisomy
A missing chromosome	Monosomy
An extra part of a chromosome	Duplication
A missing part of a chromosome	Deletion
A part of one chromosome moved to another	Translocation
A part of a chromosome turned around	Inversion

IN PURSUIT OF THE GENETIC MATERIAL

Genetics began with observations of trait transmission in families and shifted to the level of the chromosome as technology improved. The next level of genetic information is DNA. The clues to understanding the biochemical nature of the gene came from diverse places—soiled bandages, darkened diapers, bread mold, microbes, and molecular models.

IDENTIFYING THE PARTS OF DNA

The first description of the genetic material came in 1871, when Swiss biochemist Friedrich Miescher (1844–1895) developed a technique to separate nuclei from cells (Miescher 1871: 441). He collected pus on bandages—a fortuitous source, because pus is full of white blood cells. Had he chosen a bloodstain to probe, he might not have obtained much DNA, for red blood cells lack nuclei, where DNA resides. Miescher named the phosphorus-rich, acidic substance he discovered *nuclein*. The soiled bandages also yielded abundant protein, which for many years would be the prime candidate for the genetic material.

Russian-born biochemist Phoebus Levene (1869–1940) further described DNA with hundreds of experiments in his laboratory at Rockefeller University in New York City. His discovery of the sugars ribose, in 1909, and deoxyribose, in 1929, revealed two types of nucleic acid. He also discovered the DNA bases adenine (A), guanine (G), cytosine (C), and thymine (T), and confirmed the presence of phosphates. Levene deduced that DNA is built of units, which he named **nucleotides**, that consist of a single deoxyribose sugar, a phosphate group, and a base. Figure 2.4 depicts DNA at several levels, beginning with the building block, the nucleotide.

Levene got the details of DNA structure correct, but erred in envisioning the bigger picture. He claimed that a DNA molecule consists of exactly four nucleotides, one for each base, forming a structure too simple to carry genetic information. Others would consider a longer molecule, but continue the idea that the bases were equally represented, disqualifying DNA for a presumed lack of variability. Levene, like Mendel, died before his work on DNA could be appreciated, but it would prove pivotal in deducing the molecule's structure.

ONE GENE, ONE PROTEIN

Between the times that Miescher and Levene provided insight into the composition of the genetic material, a physician with a not-very-good bedside manner contributed a key clue to the function of the genetic material.

Sir Archibald Garrod (1857–1936) had a great interest in his patients' urine. Shortly before the turn of the twentieth century, as a visiting physician at a children's hospital, Garrod examined a

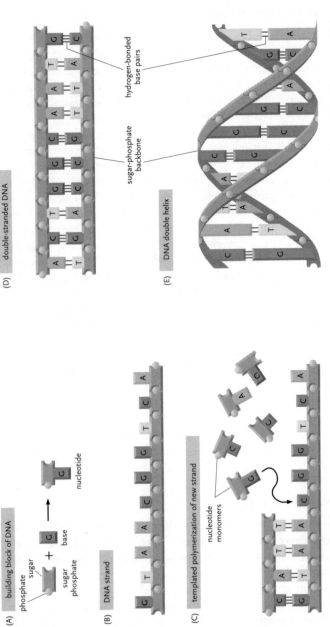

Figure 2.4 DNA and its building blocks. DNA, the hereditary molecule, is a very long molecule built of units, called nucleotides, that consist of a nitrogenous base, a phosphate group, and deoxyribose sugar

Source: *Molecular Biology of the Cell*, 5th edition, Figure 1–2, p. 3

three-month-old, Thomas P., whose diapers (nappies) had a peculiar, dark bluish-black stain. This was a sign of alkaptonuria, in which a substance called homogentisic acid builds up in joints and darkens the skin in telltale places in adults, but young Thomas appeared none the worse for his odd urine. When two siblings were born and also had black diapers, Garrod took note—not only that the parents did not have a homogentisic acid buildup, but also that they were first cousins.

Garrod had recently read Bateson's description of Mendel's work. On a walk home from the hospital one night, pieces of the mystery behind his young alkaptonuria cases began to fall into place. At the time, alkaptonuria was considered an infection, which might explain why siblings sometimes had it. But it could instead be a recessive trait, Garrod reasoned, as the parents were unaffected. His brainstorm: siblings had the condition because their parents had each contributed the same mutation, inherited by each from their mothers, who were sisters, who in turn had inherited it from the same parent. Garrod soon found other examples (Garrod 1902: 1616) and credited Mendel's laws for his epiphany. The carrier parents were the "hybrid generation," and the offspring with the darkened nappies were homozygous recessive.

Garrod was the first to link heredity to biochemistry. In general, the absence of an enzyme's activity causes the chemical that it normally breaks down to build up. Specifically, homogentisic acid builds up in alkaptonuria. Garrod named these disorders "inborn errors of metabolism." Chapter 5 tells a similar tale of foul nappies resulting from an inborn error of metabolism, phenylketonuria (PKU). Garrod died in 1936, and, like Mendel's and Levene's, his work would not be appreciated in his lifetime. In 1941, Garrod's idea resurfaced as "one gene–one enzyme," which grew out of experiments using yet another "model" organism, the bread mold *Neurospora crassa*.

George Beadle (1903–1989) and Edward Tatum (1909–1975) chose the orange-hued bread mold for the same reason that geneticists worked with fruit flies—the organism was easy to raise, and mutations were obvious (Beadle and Tatum 1941: 499). The trick was nutritional. Wild type *Neurospora* would grow on a certain medium (food) that contained simple nutrients—its metabolism would synthesize more complex nutrients. Mutants, however, would perish on this paltry diet, lacking an enzyme to build the more complex nutrients. Beadle and Tatum studied mutant mold that could only grow on supplemented medium. If, for example, mold only grew in the presence of vitamin B6, then the mutation disrupted an enzyme required for cells to synthesize vitamin B6. Beadle and Tatum concluded that one gene specifies one enzyme—which Garrod had shown, in humans, nearly 40 years earlier.

Genes are the instructions for all types of biological protein, and not just enzymes. In 1949, biochemist Linus Pauling (1901–1994) identified the protein abnormality behind sickle cell anemia (Pauling et al. 1949: 543). He was working at the California Institute of Technology (Cal Tech) on the hemoglobin molecule. Not an enzyme, hemoglobin carries oxygen in the blood. It is built of four chains of amino acids called globins, two called alpha, and two called beta (see Figure 5.2). Each globin coils around a small organic group that, in turn, shields an iron atom, creating a molecule that looks a little like four marshmallows stuck together.

Pauling and his co-workers discovered that beta globin from people with sickle cell anemia, when placed on a gel-like material subject to an electric field (a technique called electrophoresis), migrated to a different site than beta globin from a person who did not have the anemia. Furthermore, people who were carriers of the disease—inferred because their children were affected but they were not—had beta globin proteins that moved to *two* sites on the gel—the wild type position *and* the sickle cell position. Pauling had vividly unveiled the "disappearing trait" in the hybrid. At the molecular level, it didn't really disappear. In 1956, chemist Vernon Ingram added that sickle cell beta globin differs from wild type by a single DNA nucleotide. But a great deal had happened in those seven intervening years.

DETHRONING PROTEIN

The next chapter in the story of the discovery of the genetic material was to demonstrate that it is DNA, and not protein. A series of beautiful experiments led to the answer.

Frederick Griffith (1871–1941) was a medical officer at the British Ministry of Health just after World War I. He was investigating the cause of many deaths during the great influenza pandemic of 1918, which was not due to the flu virus directly, but instead to bacterial infection of the lungs. Intrigued by finding more than one strain of the bacterium *Streptococcus pneumoniae* in a single patient, Griffith wondered if the two types of bacterium could swap parts. To test his idea, he set up experiments using mice (Griffith 1928: 113).

The two strains of bacteria Griffith looked at had different outside surfaces. "Smooth" (type S) bacteria were enclosed in a carbohydrate capsule that shielded them from the mouse's immune system. They were virulent, killing mice within two days of infection. In contrast, "rough" (type R) bacteria lacked the smooth, sugary exterior and succumbed to the mouse's immune system—these bacteria were avirulent. Griffith showed that the ability to kill could be passed from type S bacteria to type R bacteria, and type R could transmit the trait when they divided.

Griffith heated type S bacteria, killing them. When injected alone, they could no longer sicken mice. But when he injected the heat-killed type S bacteria along with live type R bacteria, the mice died. Blood from the dead mice was swarming with the usually unadorned R bacteria wrapped in smooth, sugary capsules! Had something in the type S bacteria, which Griffith named the "transforming principle," entered the normally tame R bacteria, making them deadly? Again, Griffith's contribution was not appreciated in his time. He was killed while working in his laboratory in 1941, during the London blitz.

The search was on for Griffith's transforming principle. In 1944, a trio of physicians/geneticists at Rockefeller University, Oswald Avery (1877–1955), Colin MacLeod (1909–1972), and Maclyn McCarty (1911–2005), repeated Griffith's experiments, but with a twist. They added either an enzyme that destroys protein or an enzyme that destroys DNA. Type S bacteria were able to transform type R into killers when their protein was destroyed, but not when their DNA was destroyed. Therefore, the transforming principle was DNA. To confirm the results, they transferred DNA from heat-killed type S bacteria with type R bacteria into mice. Not only did the mice die of pneumonia, but their bodies yielded live type S bacteria (Avery 1944: 137).

Despite the clear evidence that DNA was the genetic material, doubt persisted, because its four-base alphabet seemed insufficiently informational in this time before combinatorial thinking. In 1951, biochemist Erwin Chargaff (1905–2002), at Columbia College, provided a key piece of information. Probing the DNA of various species, he discovered that the four bases were *not* present in equal numbers. Instead, the proportions of A and T matched, as did those of C and G.

ASSEMBLING THE DOUBLE HELIX

In the spring of 1951, 23-year-old James Watson, fresh Ph.D. in hand, was doing a year's research in Copenhagen when he heard Maurice Wilkins (1916–2004), a biophysicist at King's College in London, present his work on a technique called X-ray diffraction, which bounces X-rays off a crystal. The photographic image made by capturing the deflection of the rays reveals the crystal's shape, like sonar uses sound waves to reconstruct the shapes of sunken ships. The X-ray diffraction pattern from a hunk of calf thymus, rich in DNA, indicated a regular, repeating structure, Wilkins reported (Wilkins et al. 1953: 738).

Watson soon took a position at the Cavendish Laboratory, where he met 35-year-old Francis Crick (1916–2004). The pair set to work on determining the three-dimensional structures of proteins, but they were much more keen on DNA, especially Wilkins' X-ray diffraction views. A young physical chemist, Rosalind Franklin (1920–1958), had joined Wilkins and had produced dozens of X-ray diffraction images of DNA, in both a dry and a wet form. She accurately measured the dimensions of the molecule (Franklin and Gosling 1953: 740). However, when she presented her findings at a seminar to which Wilkins had brought Watson, she did not include the "B" form of DNA, which was swollen with water molecules.

Energized by the X-ray diffraction results, Watson and Crick asked the machine shop at the Cavendish to supply colored balls to represent atoms and sticks to represent chemical bonds, and they built models that fit the data. First, they tried a triple helix. (A single helix is impossible—it would simply unwind.) Franklin took a look and pointed out that their model was inside out—the bases should be inside the sugar–phosphate backbone.

The next clue came in the spring of 1952, when Watson discovered Chargaff's paper of a year earlier, in which he'd described the equal proportions of A and T, and G and C. When Crick read the paper, he immediately envisioned base pairs. If correct, then DNA could not only encode information, but would contain instructions and a mechanism for its duplication! The molecule would be, in essence, a self-perpetuating code of life.

Meanwhile, evidence was finally building *against* protein as the genetic material, from two researchers at the Cold Spring Harbor Laboratory on New York's Long Island, Alfred Hershey (1908–1997) and Martha Chase (1927–2003) (Hershey and Chase 1952: 39). In "blender experiments," they infected *E. coli*, a common bacterium, with a virus, T2. When a virus—a mere snippet of nucleic acid wrapped in a protein coat—infects a cell, it injects only its DNA or RNA, leaving the protein coat outside. The viral DNA or RNA then commandeers the protein synthesis machinery in the cell, instructing it to make more viruses.

Hershey and Chase grew viruses on two types of medium: one contained radioactive phosphorus, found in nucleic acids but not amino acids, and one contained radioactive sulfur, found in amino acids but not nucleic acids. Next, they infected bacterial cells. Then, they poured the material into a blender and turned it on, not creating a milkshake, but shaking free the viral protein coats clinging to the infected bacteria. Spinning the material in a centrifuge separated the viral coats from the infected cells, which sunk to the bottom of the tube. The viral coats floated to the top of the centrifuge tube and were easy to scoop off and analyze.

Where was the radioactivity in the two batches? If radioactive sulfur was in the viral protein coats but not in the infected bacterial cells, and radioactive phosphorus was in the cells but not in the protein coats, then the genetic material was DNA. This is, in fact, what Hershey and Chase saw. They had shown, definitively, that, not only is the genetic material DNA, but that it is *not* protein. The blender experiments catalyzed the race to figure out DNA's three-dimensional shape, or conformation.

At about this time, Watson saw an X-ray diffraction image taken by Franklin's graduate student that hadn't made it into her presentation the year before. "Photo 51" showed the waterlogged B form of DNA. Its X-shape indicated a helix cross section. So Watson and Crick built more molecular models, incorporating the new information: DNA is a helical-shaped acid, with a sugar–phosphate backbone on the outside and equal numbers of A and T and G and C pairs on the inside.

The machine shop could not keep up with the demands of the model builders, and so, in the morning of the last Saturday in February 1953, Watson sat at his desk, arranging and rearranging cardboard cut-outs he'd made of the bases, sugars, and phosphates. The only shape that accommodated all the experimental findings was a double helix. That made sense, Watson thought, because chromosomes duplicate shortly before cell division. He saw that arranging the base pairs side by side, rather than in short stacks as Crick had suggested, produced a molecule sleek enough to be able to wind up to fit inside a cell's nucleus. It worked!

Crick arrived within the hour, looked at Watson's cardboard creation and knew instantly that the problem was solved. The implications were profound. The elegant, symmetrical DNA double helix could do three key things that qualify it as the stuff of life: carry information, replicate, and change (Watson and Crick 1953b: 964). At lunch in the pub next door, Crick famously announced that he and Watson had discovered the "secret of life." The prestigious journal *Nature* published their paper just weeks later (Watson and Crick 1953a: 737).

REVEALING REPLICATION

Like a film whose ending suggests a sequel, the famed Watson and Crick paper concluded with, "It has not escaped our notice that the specific pairing we have postulated immediately suggests a possible copying mechanism for the genetic material." But exactly how does the DNA base sequence copy itself? In a second *Nature* paper, Watson and Crick proposed that the double helix parts, and each exposed half pulls in new nucleotides to build a new double helix—a little like a row of dancing partners separating and pulling in new mates. This mechanism is called "semi-conservative," because each double helix (row of dance partners) conserves half of itself from a previous double helix. But how could an experiment show this?

Two young researchers took up the charge to demonstrate DNA replication (Meselson and Stahl 1958: 671). In 1954, Matthew Meselson met Franklin Stahl at a summer course at Woods Hole, on Cape Cod, Massachusetts. The instructors: Watson and Crick. Meselson and Stahl met again at Cal Tech, where they would conduct what many have called "the most beautiful experiment in biology." It not only demonstrated semi-conservative DNA replication, but ruled out the other two possibilities—that one double helix creates another (conservative replication), or that a double helix shatters and rebuilds itself into two using spare parts (dispersive replication).

Meselson and Stahl marked newly replicated DNA with radioactive nitrogen in its bases to follow its fate. The radioactive DNA was denser, making it sink further than non-radioactive DNA in a centrifuge tube. By growing bacteria on radioactive medium, then switching the cells to non-radioactive medium, Meselson and Stahl deduced, from the patterns of heavy and light cells over a few generations, that DNA replication is in fact semi-conservative, and not either of the other two mechanisms. Their findings were repeated in other species, with DNA marked in different ways.

CRACKING THE GENETIC CODE

While Meselson and Stahl tackled DNA replication, others were investigating how DNA guides a cell to manufacture protein. How does a 4-letter DNA alphabet encode a 20-letter amino acid alphabet? How is DNA recycled, rather than consumed?

The first step to understanding how DNA encodes protein was to orient events in the cell. Experiments that broke open cells showed that proteins are synthesized, even when nuclei are removed, on structures called **ribosomes**. Therefore, a molecule other than DNA must carry the instructions for a protein out of the nucleus. Crick suggested that a small molecule ferries the information to the cytoplasm, where amino acid building blocks are strung together. What is this molecule? And what, exactly, *is* the genetic code?

Logic provided some clues. The first question: *How many DNA letters specify an amino acid?* Three nucleotides are the minimum. One letter obviously would not work, nor would two—A, C, T, and G can combine in only 16 ways, and 20 were required. In 1961, Crick led experiments to show the three-letter minimum using a virus called T4, whose entire base sequence was known. By treating the virus with a chemical that removes nucleotides, the researchers found that deleting one or two that were next to each other had a devastating effect on the encoded protein, but that deleting three bases in a row was not as severe.

Logic also guided experiments to answer the second question: *Is the genetic code overlapping?* If so, then certain amino acids would always follow others in a protein's sequence. When researchers sequenced several proteins, this was not the case.

Experiments answered questions too. While some researchers were identifying the three most abundant types of RNA molecule in cells—**messenger RNA** (mRNA), **ribosomal RNA** (rRNA), and **transfer RNA** (tRNA)—two young researchers at the National Cancer Institute in Bethesda, Maryland, Marshall Nirenberg and Johann Matthaei, were using a soup of ribosomes, enzymes, various RNAs, and amino acids, called a "cell-free system." They synthesized short RNA molecules that consisted solely of the RNA base uracil (chemically similar to the DNA base, thymine), and added it to the cell-free system. What would happen? Out of the soup came snippets of protein consisting entirely of the amino acid phenylalanine—and the number of phenylalanines was exactly one-third the number of RNA nucleotides, confirming the triplet nature of the code. This experiment revealed the first RNA word, or **codon**, of the RNA alphabet that is the genetic code: UUU = phenylalanine.

Repeating the experiment with different short RNAs filled in the rest of the code. First came AAA (lysine) and CCC (proline). Then came "co-polymers," such as UGUGUG . . ., which, in the test tube soup system, directed synthesis of alternating cysteines and valines, two other amino acids. But which was which? Was UGU cysteine and GUG valine, or vice versa? Analysis of a different co-polymer, UGUUUUGUUUU . . ., provided the answer. It specified phenyl-alanines alternating with cysteines. As UUU = phenylalanine, then UGU must encode cysteine, and then GUG must encode valine. In this manner, the researchers matched 64 mRNA codons to their amino acid counterparts (Crick 1961: 1227).

By 1965, the team had cracked the entire genetic code. A critical characteristic of the genetic code is that it is the same in all organisms, as well as viruses. That is, the same codons specify the same amino acids, whether in a tulip or a tiger. This "universality" of the genetic code is profound evidence that all life descends from shared ancestors.

The Genetic Code				
1st position (5 end)	2nd Position			3rd Position (3 end)
	U	C	A	G	

	U	C	A	G	
U	Phe	Ser	Tyr	Cys	U
	Phe	Ser	Tyr	Cys	C
	Leu	Ser	**STOP**	**STOP**	A
	Leu	Ser	**STOP**	Trp	G
C	Leu	Pro	His	Arg	U
	Leu	Pro	His	Arg	C
	Leu	Pro	Gln	Arg	A
	Leu	Pro	Gln	Arg	G
A	Ile	Thr	Asn	Ser	U
	Ile	Thr	Asn	Ser	C
	Ile	Thr	Lys	Arg	A
	Met	Thr	Lys	Arg	G
G	Val	Ala	Asp	Gly	U
	Val	Ala	Asp	Gly	C
	Val	Ala	Glu	Gly	A
	Val	Ala	Glu	Gly	G

Figure 2.5 The genetic code. The same mRNA codons specify the same amino acids in all life on Earth. The genetic code is therefore universal—there is no "human genetic code"

Source: Adapted from *Molecular Biology of the Cell*, 5th edition, Figure T:1

The common phrase "the human genetic code" is incorrect—what is meant is usually the "human genome sequence."

THE HUMAN GENOME PROJECT

In the 1950s and 1960s, experimental pathways converged to discover and describe the genetic material. In the 1970s, research directions diverged. Some investigators focused on the details of protein synthesis, some mapped genes to chromosomes, and others invented the biotechnologies that would eventually make sequencing the human genome feasible. Early headlines highlighted recombinant

DNA technology, which combined DNA from different species, courtesy of the universal genetic code.

THE BACKDROP: RECOMBINANT DNA, DNA SEQUENCING, AND GENE MAPPING

Careful thought preceded the first "gene splicing" experiments, which would combine DNA sequences from two types of organism in one cell. In early 1975, 140 molecular biologists met at Asilomar, California, to discuss the safety of the approach. The researchers drew up plans to contain such organisms in the lab, including ways to booby-trap their creations with genes that would kill them should they "escape." Rather than a route to producing "triple-headed purple monsters," as one researcher joked, recombinant DNA technology instead turned out to be a way to mass-produce human proteins that could be used as drugs, without needing human donors. This would become especially important in the 1980s, when AIDS arrived and made use of blood products unsafe. The first recombinant drug, human insulin, became available to people with diabetes in 1978. It had previously come from pigs or cows, and some people with diabetes were allergic to these sources of insulin. Today, a few dozen drugs are manufactured using recombinant DNA technology.

The 1970s were also the years when sequencing DNA became possible. Fredrick Sanger, at the Medical Research Council (MRC) Laboratory for Molecular Biology in Cambridge, UK, and Alan Maxam and Walter Gilbert, at Harvard University, invented different ways to do this. These approaches, and others that would follow, used the same logic: cut several copies of the same long DNA molecule into pieces that differ in size by one base from the same end, with that base labeled in some way (radioactivity, a chemical modification, or fluorescence). Aligning the pieces by size reveals the sequence, read in the order of the overhanging bases. DNA sequencing was automated by 1982 and commercialized by 1987. In the mid 1980s, a researcher could sequence up to 1,000 DNA bases a day; today, that number is in the billions. Newer techniques use microfluidics, in which small sequences of DNA are attached to tiny beads, and other variations on the nanotechnology theme.

Mapping genes to specific chromosomes also flourished in the 1970s and 1980s. Often, a gene hunt would start with an individual who had an unusual syndrome and a specific chromosome glitch, such as a missing or moved part. Linkage data were important, too. By the mid-1980s, human chromosome maps had become so crowded with genetic landmarks that several researchers realized that sequencing the human genome was within reach.

GREAT MINDS THINK ALIKE (PART 2)

This chapter has chronicled the evolution of the field of human genetics, from observations of trait patterns in families (again, peas or people), to chromosomes, and then to the level of DNA base sequences. The story continues, meeting the present time, with a reverse in direction to genomics, the study of genomes. Back in the 1970s, when sequencing just a single gene in a simple bacterium might take up an entire Ph.D. project, knowing the DNA sequence of the entire human genome was unimaginable.

The idea to sequence the human genome arose in at least three places at about the same time. Late in 1984, Robert Sinsheimer, chancellor of the University of California at Santa Cruz, assembled a group to consider sequencing the human genome as a fundraising goal. In late 1985, to the south, at the Salk Institute, Renato Dulbecco proposed sequencing the human genome to learn about the origins of cancer. The third founder, Charles DeLisi, had the power of a huge organization behind him—he was director of the US Department of Energy's (DOE) Health and Environmental Research Programs. The DOE had a long-standing interest in the effect of radiation on the human genome, because the agency developed and delivered the atomic bomb and was monitoring mutations in survivors of the blasts.

Not everyone thought sequencing the human genome was a great idea. Would such "big" science steal precious funds from "little" science, such as identifying the genes behind rare disorders? Would it divert funds from fighting AIDS? Sequencing the genome was compared to climbing Mt. Everest just because it is there. Some scientists scoffed at the value of gobs of data, compared with the traditional research route of discovering mechanisms.

In an effort reminiscent of the scrutiny of the safety of recombinant DNA research, the genetics community debated the challenges and promises of sequencing the genome. In the US, the National Research Council of the National Academy of Sciences took the lead, and both DOE and the National Institutes of Health (NIH) funded the beginnings of a 15-year human genome project in 1988. The National Human Genome Research Institute, part of the NIH, formed a year later, with James Watson as director, and the project shifted from DOE to NIH. In the UK, the MRC, Imperial Cancer Research Foundation, and the Wellcome Trust were the main players, and, in France, this role fell to a joint venture, Genethon. Similar efforts were underway in the USSR, Italy, Japan, Canada, and Latin America. The Human Genome Organization emerged to oversee the ten major sequencing centers and other academic labs and companies staking out parts of the genome.

Participants in the various projects shared concerns about the ultimate use of DNA information, especially in light of the legacy of eugenics that tainted the field of genetics earlier in the century. Five percent of the NIH budget went towards investigating ethical, legal, and social issues. Genome research throughout the world also called for the sequencing of genomes of model organisms (Table 2.5), and improving sequencing and data analysis. By 1991, companies and a few not-for-profit organizations had joined the genome sequencing effort.

Further technological advances laid the groundwork for the post-genome era, when the mountains of data would require analysis. At Stanford University in 1991, Patrick Brown invented the "gene chip," or DNA microarray. A chip reveals the mRNAs, and therefore the proteins, made in a particular cell under particular conditions, by copying the mRNA into DNA called complementary, or cDNA. Microarrays have revolutionized the study of gene expression and interaction. Also in 1991, J. Craig Venter, a biochemist at the NIH, invented expressed sequence tag technology, which is a way to pull out pieces of protein-encoding genes from a genome. The idea was perhaps an echo of the protein prejudice of years past, for nobody yet had an inkling that these sequences account for only 1 percent of the genome!

THE FINAL RACE

New computer algorithms greatly sped the aligning and reading of DNA sequences. By 1995, DNA sequencing was ten times faster than it had been at its start, two decades earlier. That was the year that

Table 2.5 The (fill-in-the-blank) genome project. Hundreds of species have had their genomes sequenced. Most are microbial, but some larger representatives are listed here

Species	Information revealed
Cat	Model for HIV research
Chicken	Influenza origin and transmission
	Clues to dinosaurs (bird ancestors)
Chimpanzee	What makes us human
Coelacanth (fish)	Conquering land habitats
Cow	How breeds differ
	How to improve meat and milk
Dog	Breeds resulted from extreme artificial selection
	Human disease model (arthritis, cancer, epilepsy, heart disease)
Honeybee	Biological underpinnings of insect societies
	Honey production
Horse	New diagnostics and treatments for racehorses
	Breeding strategies to avoid unwanted traits
Mice and rats	Have versions of nearly all genes that cause disease in humans
Mosses	Conquering land habitats
Pig	Human disease model (heart disease, obesity)
	Organ donors
Poplar	Improved wood products
Rice	Improved nutrition as dietary staple
White rot fungus	Clean hazardous waste sites

Venter and his team, now at the Institute For Genomic Research (TIGR), in Rockville, Maryland, sequenced the first genome of a free-living organism, the bacterium that causes ear infections and meningitis. At 1,830,137 bases, its genome is 1,500 times smaller than ours.

In 1998, Venter formed Celera Genomics. He vowed to beat, at less expense, the ten major sequencing centers and other groups, which had formed the International Consortium. The race grew nasty. The Consortium stressed public access to data; Celera planned to sell genome information. The two groups also differed in approach. Celera used "whole-genome shotgunning," which blasted genomes to smithereens and then assembled the sequence. The others used a "clone-by-clone" approach that hung sequenced DNA pieces onto frameworks of individual chromosome maps. The two strategies are a little like assembling a jigsaw puzzle, one random piece at a time (shotgunning) versus assembling a section at a time based on clues in the emerging surrounding picture (clone-by-clone). Shotgunning turned out to be more efficient. However, it misses repeated sequences, which are a form of genetic information too—resequencing using different techniques identifies them.

Francis Collins, currently the director of NIH, helmed the human genome project in the US, as the finish line loomed. Celera Genomics indeed had their sequence—which included DNA from Venter himself—by March 2000. Meanwhile, a new algorithm enabled the public group to finish by June. On June 26, 2000, perhaps embarrassed by the publicity surrounding the acrimony, Venter and Collins stood stiffly with President Clinton at the White House to announce the end of the race, which they graciously deemed a tie. Both groups published their results in March 2001 as a "first draft," followed by publication of the complete sequence two years later in mid April, to coincide with the 50-year anniversary of Watson and Crick's first paper (Venter et al. 2001: 1304; International Human Genome Sequencing Consortium 2001: 860).

The availability of the human genome sequence has been a boon to researchers; to clinicians, who now have more diagnostic tests for Mendelian disorders; and to drug developers, who have many new targets to pursue. Perhaps unanticipated is the use of the sequence to unveil normal variation, the differences and distinctions that make us each unique. That analysis will continue for some time to come, because we now know that there is not "a" human genome, but many variations—where the field of genetics actually started.

Looking back, did the sequencing of the human genome really reveal what Francis Collins had called "the molecular essence of humankind?" Perhaps not, for it is not the blueprint itself that defines an individual, but how that blueprint is used, in an ever-changing panoply of situations.

Francis Crick glimpsed the complexity that we would encounter in the genome by seeking to learn our genetic blueprints back in 1966, when he pondered what could be learned from sequencing the genome of E. coli. Of the thousands of genes in E. coli, he predicted in *Molecules and Man,* we would need to learn what each does and how their actions are controlled. He wrote, "this particular problem will keep very many scientists busy for a long time to come."

How very true!

WHAT WOULD YOU DO?

Species other than humans have been essential in genetics research. Gregor Mendel worked out the two laws of inheritance using pea plants. The earliest views of chromosomes came courtesy of salamanders, grasshoppers, and sea urchins. The first genetic map represented thousands of fruit fly lives. Finally, deciphering DNA's structure and revealing its function depended upon experiments using viruses, bacteria, bread mold, and others. In short, we could never have sequenced the human genome without the contributions of a diversity of other organisms.

Many people become upset at the idea of using animals in biomedical research. Animal rights organizations attempt to improve the situation for animals used in research, or object to it entirely (People for the Ethical Treatment of Animals; www.peta.org/). Scientific organizations take the opposite viewpoint (Americans for Medical Progress, www.amprogress. org/). The American Society for the Prevention of Cruelty to Animals (www.aspca.org) takes the middle road, stating that, "animals should be used to develop medical treatments only when alternatives do not exist, and when the research goal is of very significant humanitarian value." However, the very nature of science is to test a hypothesis—the outcome is not known, nor does basic research have a goal other than to answer a specific question.

Another aspect of the animal research dilemma is "the Bambi effect"— that is, people tend to object more to research that uses mammals than to research that uses other types of animal. The author of this book is guilty of this inconsistency—as a graduate student she killed possibly millions of fruit flies, but never touched a mouse, rat, or rabbit.

How do you feel about the use of animals in research? Did reading this chapter change how you feel about the issue? List factors that you think are important in deciding if animals should be used in an experiment. Would you feel differently about a particular experiment that is conducted on eels or cockroaches versus kittens? What is an alternative to using animals in research?

SUGGESTED READING

The chapter cites the key papers in the history of genetics, but many books also recount the experiments and events that led to our understanding of the genetic material. Henig (2000) and Orel (1996) cover Mendel's life and experiments. Sturtevant (1965) provides peeks into the Fly Room from the inventor of genetic maps. Crick (1966), Hoagland (1990), Keller (2000), McElheny (2003), and Watson (1968; 2003) discuss the middle years of the quest. Shreeve (2004), Davies (2002), and McElheny (2010) relate the race to sequence the human genome, and historical information on the project can be found on the websites of the two US government agencies involved from the start: the NHGRI, at http://genome.gov, and the DOE, at http://genomics.energy.gov. Human genome information from the UK can be found at the Wellcome Trust Sanger Institute (www.sanger.ac.uk/).

3

HOW GENES WORK

Inherited traits and illnesses affect us at several levels. A molecular glitch disrupts how a certain type of cell functions, causing the tissue and organs of which those cells are a part to also malfunction. Symptoms result. How genes are distributed as sperm and egg form and then come together at fertilization explains familial patterns of trait transmission. How we choose our partners explains much of our evolution.

Living with an inherited disease hits all the levels of genetics, as a mutation affects an individual and reverberates through a family and perhaps a population. This chapter examines the levels of genetics through the lens of a single-gene disease, cystic fibrosis (CF) (Table 3.1).

MIKAYLA'S STORY

"We found out when I was about five months pregnant that Mikayla would have CF," recalls Donna Polski of Minneapolis, Minnesota, whose daughter was born on November 19, 2002.

I was having amniocentesis because I was over 35, looking for conditions related to the mother's age. The genetic counselor suggested CF testing because of our northern European background. We didn't have CF in the

Table 3.1 Levels of genetics

Biological level	Cystic fibrosis as an example
Molecular	A protein that forms a channel for salt components in cell membranes is destroyed or does not function.
Cell	Cells lining the airways, ducts in the pancreas, and tubes that carry sperm, and cells of salivary and sweat glands require the protein to secrete.
Tissue	Lining cells (epithelium) and secretory cells are affected.
Organ	The lungs, pancreas, and skin are affected.
Individual	A person experiences "failure to thrive" and respiratory and digestive problems.
Family	CF is inherited as an autosomal recessive condition.
Population	The geographical distribution of CF today reflects protection against an infectious disease.

family, so we weren't sure we really needed to take the test. Although recommended, it wasn't yet routine to test for CF in pregnancy. The insurance company approved, so I had a blood draw to "rule out" being a carrier, and was shocked to find out that the test came back positive. Next, my husband had a blood test that identified he was a carrier too. So the lab doing the amnio checked the amniotic fluid, and the next day, we got the phone call that our baby would have CF. We were really shocked.

Donna and her husband Michael were right to be surprised—CF affects only 1 in every 2,500 non-Hispanic Caucasian newborns in the US, where 30,000 people have CF. The global figure is about 70,000, and millions of people are carriers. The disease upsets the transport of chloride (Cl^-) and other ions (charged atoms and molecules) into and out of cells through specialized channels. This alters secretions in ways that clog narrow passageways in several body parts, most often affecting breathing and digestion. Mikayla inherited two copies of the most common mutation, called *delta F508*, known in the CF community as a "double delta" genotype. Donna and Michael each have one copy of the *delta F508* allele. Double deltas typically have severe symptoms.

Mikayla's medical team acted right away, knowing that the pancreas and lungs can begin shutting down before birth. Ultrasound scans of Donna's uterus already showed a blockage in Mikayla's intestines, seen in one in five people with CF. Doctors delivered Mikayla early, and surgery was performed the next day to open up her intestine, but it ruptured and had to be repaired. She spent her first month in the neonatal intensive care unit. Donna recalls,

> They started big guns antibiotics right away, and pancreatic enzymes to help her digest fats. She came home mid-December. Two weeks later, she developed aspiration pneumonia and MRSA (methicillin-resistant *Staphylococcus aureus* infection), and so she was hospitalized for another month, and given intravenous antibiotics.

The hospital was to become a familiar place for the little girl, who contracted a parade of viral and bacterial infections. By age six, the visits became much less frequent, perhaps because of the daily "bronchial drainage" treatments Donna and Michael began as soon as Mikayla came home. They would gently tap on their daughter's lungs, in different positions, to loosen the sticky mucus that is the hallmark of the disease. By the time she was 16 months old, Mikayla had grown enough to switch to using a vibrating vest.

Today Mikayla follows a complicated daily routine (Figure 3.1). At night, a feeding tube provides extra calories. After she awakens, she dons the vibrating vest for 30 minutes, while using her laptop or watching TV. At the same time, she uses a nebulizer to inhale antibiotics, bronchodilators, mucolytics, steroids, and hypertonic saline, delivering the drugs directly to her lungs. She receives pancreatic enzymes through the feeding tube, or Donna opens pills and shakes the enzymes onto a soft food such as applesauce. These enzymes are vital for digesting the fat-soluble vitamins. "We try to pack in the calories. She gets lots of butter, whipped cream, and high fat foods," says Donna. Mikayla goes to school, and then repeats the treatment regimen before dinner. When she is ill, Mikayla does her "vest/nebs" up to four times a day.

As Mikayla grows up, she is taking over more of her care. For now, what bothers her most is being unable to play with her friends, or go to parties, in an effort to avoid infection. Life with CF means constant awareness of pathogens. Donna says,

Figure 3.1 Mikayla Polski giving herself a treatment. Mikayla has cystic fibrosis. At least twice a day, she gives herself a nebulizer treatment, which delivers drugs directly to her clogged lungs

> Did you know that *Pseudomonas* bacteria live in soil and stagnant water, like old hoses? We let Mikayla run through the sprinkler, but only if the hose is new. At pools, if the chlorine is too high, it affects her lungs, but if it is too low, she can get *Pseudomonas* or mold in her lungs. We have to be careful around bonfires, because of the smoke and particulates.

The dedication to Mikayla's care has paid off, for she has made enormous strides. "I almost feel like a normal kid. I am having so much fun," she said in early 2010. When Mike and Donna asked her what she meant, the little girl explained that, because she can now attend school, and take skating and swimming lessons, she sometimes forgets that she has CF! She's even ridden a horse and gone jet skiing. But her father, Mike, recalls when the situation was much more dire. "There were times we didn't know if she would make it through the night—but she fought and she did."

With luck and perseverance in preventing symptoms, Mikayla may live several decades. This wasn't always so. In 1955, most youngsters with CF didn't live long enough to enter kindergarten;

today, median survival age is 37. Those born after 2000 are projected to live until age 50 or beyond. Survival has been steadily increasing, as researchers devise ways to prevent and attack the symptoms. Still, young people die of CF. Laura Rothenberg chronicled her battle with the disease on a radio program in the US (National Public Radio's *All Things Considered*), called "My so-called lungs." She died following a double lung transplant in 2003, at age 22.

Before discovery of the CF gene, mutations, and malfunctioning protein, which made the Polski family's testing possible, the first signs of illness were, typically, severe respiratory infections and a vague "failure to thrive." Diagnosis could take months or years. For example, 52-year-old "Mr. Y" recalled how his parents had noted his large, greasy stools when he was an infant, but he was not diagnosed, with a traditional "sweat test" to detect the chloride defect, until a bout with pneumonia at age three (Boyle 2007: 1787). Today's early detection of CF extends lives (Kerem et al. 1989: 1073).

INSIDE THE CELL

The story of Mikayla's CF begins with a gene in the middle of the long "arm" of the seventh largest chromosome. Sprinkled among the gene's 250,000 DNA bases are 6,129 bases, distributed in 27 segments, that encode the 1,480 amino acids of the cystic fibrosis transmembrane conductance regulator protein, or CFTR. On both copies of Mikayla's chromosome 7, the *CFTR* gene is missing the same three DNA bases in a row, which encode the 508th amino acid in the sequence of 1,480. The name of her mutation—*delta F508*—means "deleted at position 508." (Italics denote a gene; roman letters denote a protein.)

Mikayla's double delta mutations are in each of the trillions of cells of her body, but cause her symptoms by affecting specific types of cell that use CFTR protein. Each cell of the body (except the sex cells) contains two copies of the genome, but different cell types actively use different subsets of the genes.

MAKING A PROTEIN

Imagine a cell lining a bronchial tube leading to one of Mikayla's lungs. The cell arose when another lining cell divided, replicating its

DNA and passing on the faulty instructions missing the crucial trio of bases in the *CFTR* gene. Some of the DNA is transcribed into messenger RNA (mRNA), which carries the information to synthesize proteins that the cell needs, both to stay alive and to perform specific functions, such as regulating the composition of secretions.

The mRNA for CFTR is snipped and stitched into its final 6,129 nucleotides. Then, it exits the nucleus through a nuclear pore, entering the gel-like **cytoplasm** where structures called **organelles** compartmentalize the cell (Figure 3.2). Some organelles function as molecular machines, with specific functions. The mitochondrion, for example, houses the energy reactions of the cell. Some organelles sequester biochemicals that might harm other cell parts. A lysosome, for example, is a sac of enzymes that dismantle debris.

Several types of organelle form a protein-processing network. It folds proteins, adds sugars or fats to some of them, and then either

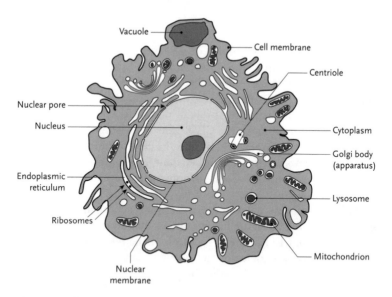

Figure 3.2 The anatomy of a cell. Organelles are specialized structures that divide the work of the cell, increasing efficiency and protecting delicate cell parts

Source: Adapted from *Applied Genetics in Healthcare*, Figure 2.15, p. 28

sends a protein to function inside the cell, or packages it in an oily bubble for jettisoning from the cell as a secretion. The secretory network consists of the endoplasmic reticulum (ER), ribosomes, and a Golgi body. The same complementary base pairing that lies behind DNA replication, discussed in Chapter 2, also guides protein synthesis. A binds with T (in DNA) and U (in RNA), and G binds with C.

The manufacture of CFTR protein begins as its mRNA exits the nucleus and contacts ribosomes that stud the ribbon-like ER. Ribosomes, made of ribosomal RNA (rRNA) and proteins, are staging areas for the assembly of a protein. First, an rRNA pairs with a few bases at the start of an mRNA in a way that exposes the mRNA's entire base sequence. Then, molecules of a third type of RNA, transfer RNA (tRNA), form complementary base pairs with the mRNA. Transfer RNAs are small, fold into a cloverleaf shape, and come in several different varieties, each type bonded to one of the 20 types of amino acid. The key to tRNA's role as a "connector" is its specificity: each of three bases in a row in the mRNA—a codon—attracts the complementary three bases of part of a tRNA called its anticodon. A tRNA with a particular anticodon always carries the same type of amino acid. Figure 3.3 is an overview of protein synthesis, a process called translation.

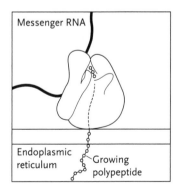

Figure 3.3 Protein synthesis. The DNA sequence of a gene includes the information for a specific sequence of amino acids

Source: Adapted from *Introducing Genetics: From Mendel to Molecule*, Figure 12.7, p. 176

The choreography among the RNA types and the organelles with which they interact operates a little like an appliance coming off an assembly line. The protein must not only have the right components in the right order, but it must fold into a specific three-dimensional form, or conformation. To follow this complex process, consider the section of CFTR mRNA that is the site of Mikayla's *delta F508* mutation:

A U C U U U G G U

In the normal (wild type) *CFTR* allele, these three mRNA codons attract tRNAs that bring in the amino acids isoleucine (AUC), phenylalanine (UUU), and glycine (GGU). The *delta F508* mutation lacks the UUU. Normally, ribosomes deliver wild type CFTR into the labyrinth of the ER, where the proteins move in the winding channels towards the cell's outer surface (cell membrane). In Mikayla's cells, this transport halts, because the one absent amino acid disrupts the folding of CFTR proteins into their final form. This is a major glitch.

PROTEIN FOLDING

Proteins must fold a certain way to be active. In a way, proteins police themselves. Proteins called chaperones peruse forming proteins, detecting misfolded local regions and correcting them, like a parent gently helping a child navigate an arm through a sleeve. If misfolding persists, another protein, called ubiquitin, attaches and herds the doomed protein towards a spool-like structure called a proteasome. The errant protein is threaded through the proteasome and, like food being pushed through a garbage disposal, is torn into pieces. In this way, delta F508 CFTR protein is dismantled before it even escapes the ER to reach the Golgi body. This fate befalls all of Mikayla's CFTR proteins, and half of her parents'. Perhaps, one day, she and others will take "corrector" drugs that rescue the misfolded proteins and escort them to the cell membrane, restoring working ion channels. Correctors are in clinical trials, but existing drugs may help correct protein folding too. One is the erectile dysfunction drug sildenafil (Viagra) (Dormer et al. 2005: 55).

Table 3.2 The events of protein synthesis: Making CFTR

1. Cell receives signals to produce CFTR proteins.
2. *CFTR* gene is transcribed into mRNA, which is cut into a final form and traverses nuclear pores into the cytoplasm.
3. CFTR mRNAs bind ribosomes. tRNAs bring in the appropriate amino acids, building CFTR protein, which enters the ER.
4. Ubiquitin yanks misfolded CFTR proteins from the ER and sends them to proteasomes for destruction.
5. Normal CFTR proteins continue on to Golgi body, picking up sugars.
6. CFTR proteins are shuttled in oily bubbles to the cell membrane, where they embed and form ion channels when ATP binds and splits.

A normal CFTR protein continues through the ER, emerges in a bubble, and glides to the next stop, the glistening stacks of the Golgi body. Here, the protein picks up sugars, for CFTR is actually a *glyco*protein. Sugared CFTR protein is then shuttled to the cell membrane, where it nestles into the double lipid layer. When the protein binds ATP, the biological energy molecule, it contorts, forming a channel, like a lump of dough being shaped into a doughnut. Then ATP splits, releasing energy that powers the transit of chloride ions into and out of the cell. CFTR also regulates other ions, including sodium (Na^+) and bicarbonate (HCO_3^{-1}). Table 3.2 summarizes the cellular events that go awry in cystic fibrosis.

Because a gene carries information, it can vary, just as letters in a sentence vary when typographical errors occur. More than 1,600 mutations riddle the *CFTR* gene. Also like the letters in a sentence, some of these genetic changes cause symptoms, but, for most, the "meaning" is clear, and the person is healthy or has mild symptoms. Mutations can disrupt a protein in several ways: altering the amino acid sequence, adding sugars or fats, or changing production rate, transport, or stability.

THE INDIVIDUAL: TISSUES AND ORGANS

Mutations affect cells, which affect the tissues and organs of which they are a part, producing the distinctive signs and symptoms of a genetic syndrome. For some conditions, the connection between mutation and health is obvious. A person missing a clotting factor gene bleeds and bruises easily. A boy with muscular dystrophy, missing a protein his muscle cells need to withstand contraction, grows weaker.

THE PHENOTYPE

When the signs and symptoms of an illness—the mutant phenotype —reflect the same underlying problem in different body parts, the condition is termed **pleiotropic**. CF is a classic example, because abnormal CFTR protein affects the composition of several body fluids, including sweat, saliva, mucus, tears, and digestive enzymes.

The most obvious symptoms of CF affect the respiratory system. The altered concentrations of various ions in the mucus that lines the tubes and tiny air sacs of the respiratory tract provide a welcoming environment for bacteria that rarely colonize healthy lungs— *Pseudomonas aeruginosa*. Nearly 80 percent of CF patients have this infection by age 18. It produces shortness of breath, fatigue, a productive cough, and decreased lung function, but no fever. During *P. aeruginosa* infection, fluid in the lungs becomes acidic, and inflammation ensues as the immune system sends in armies of white blood cells, which release leukotriene B. This substance beckons more inflammatory cells, which secrete proteases, enzymes that destroy airway lining cells. To temper the inflammation, people with CF take anti-inflammatory drugs, as well as a drug called Pulmozyme that breaks down DNA from accumulated white blood cells. Inhaling hypertonic saline—a saltier solution than is inside airway lining cells —draws water out of the cells, loosening secretions. Aerosols of antibiotics combat infections. More than 60 species of bacteria can reside in the sinuses, airways, and lungs of CF patients.

Exercises to loosen mucus are part of the daily routine of a person with CF. The hair-like cilia that emanate from cells lining the airways normally sweep mucus up and out of the body. Not so in CF. Several "airway clearance techniques" dislodge the mucus, so that the person can spit it out. Often, caregivers pat and knead the chest to shake free the mucus, as Mikayla's parents did when she was an infant. Devices also help a person breathe out forcefully in a way that sends air beneath the mucus layer to break it up. Other respiratory problems common in CF are sinus infections, nasal polyps, bronchitis, and bronchiectasis (dilated, inflamed, and collapsed bronchial tubes).

Nearly all people with CF also feel the effects of an impaired pancreas, which is responsible for the vague diagnosis of "failure to thrive." Blocked delivery of digestive enzymes to the small intestine causes vitamin deficiencies, malnutrition, abdominal pain, and fatty, greasy bowel movements. Most people also have salty sweat from the abnormal ion transport in sweat glands. CF may affect fertility. In about 99 percent of men with CF, thick mucus blocks the sperm-carrying tubes, which degenerate. For some genotypes, this is the only symptom. In women, CF thickens the secretions around the cervix.

OUTSIDE INFLUENCES

Inheriting two mutations in the *CFTR* gene can cause CF, but the severity, rate of progression, and other aspects of an individual's experience depend also upon the actions of other genes and environmental influences. This makes sense. A child with CF will have more severe respiratory problems if exposed to air pollution or pathogens. For example, the 2009 H1N1 pandemic influenza made most people only mildly ill, but could be deadly to those with CF. The environmental component can empower people with CF to use exercise to ease symptoms—if they can find time. One young woman takes medications and dons her vibrating vest four times a day, for an hour each time, but also manages to swim and run. Another young woman finally began to have days without feeling as if she had a severe cold when she began running—"with every step, I'm beating CF!," she says.

Other genes affect the symptoms and severity of CF. Meconium ileus, the blocked intestine discovered in Mikayla before she was born, occurs if an individual inherits a mutation in a gene called *CFM1* along with *CFTR* mutations. Mutation in a gene whose protein product interacts with CFTR can cause CF-like symptoms. This is the case for a protein called filamin A, which attaches CFTR proteins in the cell membrane to the protein network beneath it.

THE FAMILY

Remember Mendel? The laws he derived from following the inheritance of several traits in peas whose genes are on different chromosomes apply to any species whose cells have two copies of each chromosome. Tall plants that, when crossed, give rise to tall and short offspring in the ratio 3:1 illustrate the same pattern of trait transmission as Donna and Michael Polski, who each carry a *CFTR* mutation. If Donna and Michael were able to have many offspring, as pea plants do, the chances are that about a fourth of them would inherit CF, according to Mendel's law of segregation. Each child they conceive faces a one in four chance of inheriting CF, like Mikayla, a one in two chance of being a carrier, and a one in four chance of being wild type.

INHERITANCE PATTERNS

The inheritance pattern seen among the tall and short pea plants, and in a family with CF, is **autosomal recessive**. "Autosomal" refers to the location of the gene on a non-sex chromosome, or **autosome**, which does not carry genes that determine sex. Recessive refers to an allele (gene variant) that is only expressed when inherited in two copies, one from each parent. In a small family, a homozygous recessive individual may not appear for several generations, as carriers silently pass on the mutant allele. An autosomal recessive disease or trait may be more obvious in larger families. One woman, for example, was diagnosed with mild CF as a college student, when an astute physician tested her upon learning of her history of frequent sinusitis and bronchitis. She convinced her six siblings to be tested, and three of them had CF!

Geneticists use a diagram called a pedigree to trace traits and illnesses in families (Figure 3.4). It uses squares to represent males and circles to represent females, with lines to show the relationships of family members. Heterozygotes are depicted as half-filled-in shapes, and individuals who express a mutant phenotype as filled-in shapes.

Autosomal recessive is one mode of inheritance, which is the pattern of trait transmission (Table 3.3). A second mode of inheritance is **autosomal dominant**, in which only one mutant allele need be inherited to express the trait. The form of dwarfism described in Chapter 2, achondroplasia, is autosomal dominant.

Traits inherited from genes on the sex chromosomes are termed X-linked or Y-linked. A female has two X chromosomes, and a male

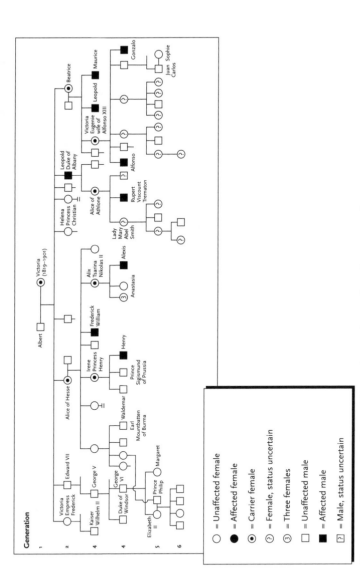

Figure 3.4 A famous pedigree. This pedigree shows the inheritance of the clotting disorder, hemophilia, from Queen Victoria of the British royal family to other royal families

Source: *Introducing Genetics: From Mendel to Molecule*, Figure 6.11, p. 83

Table 3.3 Modes of inheritance

Mode	Description
Autosomal recessive	An allele on a non-sex chromosome must be present in two copies to affect the phenotype.
Autosomal dominant	An allele on a non-sex chromosome must be present in one copy to affect the phenotype.
X-linked recessive	A male expresses a trait whose gene is on the X chromosome because he does not have a second allele to mask it; a female requires two copies of the allele to express the trait.

has one X chromosome and one Y chromosome. A male inherits an X-linked trait or condition from his mother, and his Y chromosome from his father. The bleeding disorder hemophilia A is a classic example of an X-linked recessive condition. A son who inherits an X-linked recessive mutation is affected because he does not have a second X chromosome to mask the mutant allele. The white-eyed male fruit flies that T.H. Morgan observed in the Columbia University Fly Room, discussed in Chapter 2, inherited an X-linked recessive mutation from their mothers.

X-linked dominant conditions are quite rare and are so severe that affected males do not survive to be born, and females are severely affected. An example is incontinentia pigmenti. It causes swirls of skin pigmentation and other symptoms in females, but males do not survive to be born. Y-linkage is rarely seen, because the Y chromosome has few genes.

Whether an allele is dominant or recessive depends somewhat on technology—that is, what we can measure or detect. A carrier of an inborn error of metabolism may be healthy but have only half the normal amount of the biochemical that the implicated enzyme would normally break down. Sometimes, a carrier has mild symptoms of the associated illness, a situation called incomplete dominance. This is the case for CF. Carriers are prone to asthma, perform below average on lung function tests, and make half the normal amount of sweat (Dahl et al. 2005: 1911). For some diseases, carriers have symptoms unlike those of the recessive disease. This is the case for Gaucher disease, an inborn error that causes bone pain and other symptoms. Carriers are significantly overrepresented among people with Parkinson's disease, a neurological condition.

FROM GENOTYPE TO PHENOTYPE

Phenotypes of the same disease can vary, even within a family (Table 3.4). Two terms describe this variability: **penetrance** and **expressivity**. Penetrance is an all-or-none characteristic, referring to the percentage of people with a particular genotype who express the associated phenotype. For a pleiotropic (multi-symptom) condition such as CF, penetrance is stratified by symptom. It is high, for example, for impaired breathing and digestion, meaning that most people with CF have these symptoms. Less common manifestations, such as diabetes and blocked intestines, have lower penetrance.

Expressivity refers to the severity of a symptom. It is low in a person with CF who has occasional bronchitis, but high in someone who is often hospitalized with severe lung infections. Nearly all single-gene traits and conditions are "incompletely penetrant and variably expressive," which means that not all the people who inherit a particular genotype develop the associated disease, and, for those who do, severity varies. Add pleiotropy, and single-gene inheritance in a family can be confusing, as different individuals have different experiences. CF may touch a large family as a child frequently hospitalized with pneumonia, a sibling with sinus infections, and a male cousin with infertility—they might not realize they have the same disease. In fact, CF's pleiotropy is why it took clinicians so long to fit the symptoms into a coherent story.

Table 3.4 Aspects of gene expression

Term	Definition
Penetrance	The percentage of individuals with the same genotype who express the phenotype.
Expressivity	Variability in symptom severity among people with the same genotype.
Pleiotropy	The association of several symptoms with a genotype.

A BRIEF HISTORY OF CF

Centuries ago, people recognized that a child with salty sweat was likely to waste away and suffocate. In 1596, a professor of anatomy in Leiden, the Netherlands, noticed that one such "bewitched" child had an enlarged, white, swollen, and hard pancreas. Parents of the salty children also commented on their offsprings' fatty, foul stools.

It wasn't until the 1930s that this peculiar combination of symptoms was named—"cystic fibromatosis with bronchiectasis" and "cystic fibrosis of the pancreas." In 1944, Dorothy Andersen, a physician at Babies Hospital of Columbia Presbyterian Medical School in New York City, deduced that the condition is autosomal recessive (Andersen 1944: 100). Five years later, when a sizzling summer's end caused heat prostration among her patients with "CF of the pancreas," a young colleague, Paul di Sant'Agnese, examined the suffering children and discovered three to five times the normal amount of salt on their skin (Kessler and Andersen 1951:648; di Sant'Agnese et al. 1953: 549). The old idea of predicting doom when a kiss to a child's forehead tasted salty now made sense, and researchers tried to turn the folklore into a diagnostic test. First, they put patients in big bags to get them to sweat enough to measure the salt, but this could raise temperatures to deadly levels. Finally, the "sweat test" was perfected. In it, a chemical that induces sweating is applied to a small area of skin on an arm, and an electrode delivers a weak electrical current to the area, which stimulates sweating. A warm or tingly feeling ensues for about five minutes. Then, sweat is collected for 30 minutes, on a piece of filter paper or gauze, and sent to a laboratory for analysis.

In the 1960s and 1970s, researchers closed in on a unifying factor behind the symptoms of CF. In these days before gene searches, secretions and excretions from patients—saliva, sweat, mucus, tears, and stools—were applied to all sorts of experimental systems, from cilia-fringed oyster cells to rabbit tracheas, from rat salivary glands to cells from an obscure marine worm used as fish bait (see Table 2.5). Results suggested that something about CF disrupts what goes into and out of cells that secrete—and it wasn't just a disorder of mucus, because sweat, tear, and salivary glands don't make the sticky substance.

The answer came in 1981, when researchers found the telltale salt disturbance in patients' nasal passages (Knowles et al. 1981: 1489). The culprits were electrolytes, such as table salt (NaCl), dissolved in the water of body fluids as its component ions. Table salt, or sodium chloride, is made up of equal numbers of sodium (Na^+) and chloride (Cl^-) ions. Discovery of the CFTR ion channel a few years later would show that, in CF, Na^+ enters cells too readily, and Cl^- not readily enough.

Suddenly, the spectrum of symptoms made sense: electrolyte imbalances set up a chain reaction by altering the pH (acidity/ alkalinity) inside and outside lining and secretory cells in certain places in the body. As a result, along the tubes and spaces of the respiratory system, mucin proteins are not dotted properly with sugars, while cells draw water from the secretions. The mucus changes chemically and thickens, beginning in the bronchial tubes and spreading downward into the lungs. Over time, the bronchi widen and become scarred and infected. In the pancreas, the chain reaction dries digestive juices, cutting off enzyme delivery to the small intestine. The blocked enzymes begin to digest the pancreas itself, while tiny cysts and fats infiltrate the intestine, blocking the folds-within-folds where digested nutrients normally enter the bloodstream. Malnutrition results. In about one in ten newborns, fat, mucus, and undigested proteins plug the small intestine—this is the meconium ileus that Mikayla had. Sweat is salty, saliva stringy. In some patients, the gallbladder fills with a green, jello-like substance.

Doctors discovered the various blockages of CF during autopsies, but finding the gene finally explained the pathology. The gene mapping that became possible in the 1980s narrowed down the search to about 2 million DNA bases on the long arm of the seventh largest chromosome (Knowlton et al. 1985: 380; Tsui et al. 1985: 1054). A heated race ensued between research groups in London and Toronto, culminating with the discovery of the *delta F508 CFTR* mutation in cells from the sweat glands of patients (Riordan et al. 1989: 1066). Researchers have been identifying and describing variants of the gene ever since.

THE POPULATION

The level of the population is where genetics intersects history, anthropology, and human behavior. As Chapter 1 describes, the distributions of alleles in different societies hold telling clues to the narratives of humankind, past, present, and future.

HOW POPULATIONS CHANGE

At the population level, genetic information takes the form of allele frequencies. All of the alleles in a population constitute the **gene pool**. Because allele frequencies are almost always changing, gene pools are in constant flux. When two populations change enough genetically that they can no longer interbreed, a new species arises.

Five processes change allele frequencies: mutation, migration, genetic drift, non-random mating, and natural selection (Table 3.5). These factors are not equally important. Mutation, for example, is very rare. More commonly, migrants introduce novel alleles into populations. A city such as New York or London is continually being remade, in a genetic sense, as people bring their distinctive alleles from different parts of the world.

In **genetic drift**, part of a population becomes genetically distinct when its members reproduce only among themselves. This happens when a few people leave one settlement to begin another, called a "founder effect." The inhabitants of the Faroe Islands of Denmark reveal a classic founder effect—all people with CF are "double deltas," like Mikayla, and are possibly all descended from the same individual. A founder effect need not involve physical removal—it can also

Table 3.5 Forces behind population change (evolution)

Force	Definition
Mutation	A new allele forms, such as a DNA base substitution.
Migration	Individuals bring new alleles into a population, or remove alleles.
Genetic drift	A subpopulation sequesters certain allele combinations from the larger group.
Non-random mating	Some individuals contribute more alleles to the next generation than others.
Natural selection	A phenotype confers a reproductive advantage or disadvantage.

happen when people marry only certain types of people, such as members of religious communities within a large city.

A type of genetic drift called a population bottleneck occurs when a disaster kills most members of a population, leaving a few behind to replenish the gene pool. The genetic structure of the new group is unlikely to be exactly the same as the first. A global population bottleneck is a classic science fiction theme. To escape a doomed Earth, a handful of people board spaceships, or hide in huge arks or underground, until the environmental upheaval abates. Table 1.5 lists science fiction plots that borrow from population genetics.

In non-random mating, one individual contributes more to the next generation than others by having many offspring. This also changes allele frequencies. Non-random mating is actually a goal in agriculture, where one genetic variety of crop or one bull with valuable sperm becomes the sole source of genes, to perpetuate traits that are valuable to us.

Natural selection is the most powerful force that molds population genetics, and therefore evolution, which is why it is the subject of Charles Darwin's *Origin of the Species*. Natural selection refers to the increased representation, in a population, of an inherited trait that makes an individual more likely to successfully reproduce. Darwin's phrase "survival of the fittest" actually has nothing to do with physical prowess, but instead refers to reproductive success.

WHY CF IS COMMON IN SOME PLACES, RARE IN OTHERS

When Donna Polski told the genetic counselor that her and Michael's families came from Norway, Sweden, and Finland, the counselor was alerted. She knew that both the incidence (number of new cases each year) and prevalence (percentage of affected individuals in the population at a given time) of CF are highest in northern Europe, where about 2 percent of the population carries a mutant CF allele. In fact, CF is much more common in some parts of the world than others (Table 3.6).

The principles of population genetics explain why CF and other inherited diseases affect different populations to different degrees. The language of population genetics is not the four letters of the DNA alphabet, but allele frequencies. Consider allele frequencies that change in one geographic direction, such as from west to east. These

Table 3.6 Incidence of CF in selected populations

Population	Incidence
Albania	1/555
Australia	1/2,021
Basque	1/4,500
China	1/90,000
Faroe Islands	1/1,775
Iceland	1/8,344
India	1/40,000
Italy	1/4,238
Japan	1/350,000
Jordan	1/2,560
Sweden	1/5,600
UK	1/2,415
US:	
Caucasian non-Hispanic	1/2,500
Hispanic	1/13,500
African-American	1/15,100
Asian-American	1/35,000
Ashkenazi Jewish	1/2,271

changes reflect long-ago migrations and where and when people stopped and had children with residents, introducing new alleles and diluting the frequencies of existing alleles in the established community. This directional change in allele frequency is called a **cline**. The *delta F508* CF allele follows a clear cline, falling in frequency from northwest to southeast in Europe, parallel to known migration patterns.

Once a disease-causing allele is introduced into a population, its specific effect influences whether it increases or decreases in frequency. Why, researchers have long wondered, is CF much more common among Europeans and those of European ancestry? The reason is not due to genetics at all, but to protection against an infectious disease—tuberculosis (TB). The strongest clue to CF's prevalence is the predominance of one particular mutation, *delta F508* (Table 3.7).

Why does *delta F508* predominate? It is unlikely that the exact same allele would arise anew from mutation, repeatedly. Nor could a founder effect account for the gradual nature of the cline. A more

Table 3.7 Percent of CF mutations that are *delta F508*

Population	Percentage of mutations that are *delta F508*
Worldwide	66
Brazil	23
Ecuador	25
Faroe Islands (Denmark)	100
Iceland	57
Israel	30
Northern Europe	75–88
Romania	65
Slovakia	37
US:	
Caucasian, non-Hispanic	72
Hispanic	54
African-American	44
Asian-American	39
Ashkenazi Jewish	31

plausible explanation is a variation of natural selection called **balancing selection**, in which carriers are better able to survive because they resist a second illness. The better health of the carriers keeps the genetic disease-causing allele in the population.

The classic example of balancing selection is sickle cell disease and malaria, an infectious disease that causes cycles of high fever as the pathogen curls up inside human red blood cells. Sickle-shaped cells keep the pathogens out, and carriers apparently have enough of the spiky cells to keep malaria away, but not enough to disrupt circulation, as in sickle cell disease. Geographical maps of where sickle cell disease and malaria are most prevalent coincide, in Africa, revealing the balancing selection. Sickle cell carriers are healthier than others in places where malaria kills, and have more children, perpetuating the mutation. Chapter 5 discusses sickle cell disease further.

How does CF protect against TB? The balancing selection arises from the altered processing of an enzyme called arylsulfatase B. In CF, the abnormal ion channels change the acidity in the cell in a way that affects folding of the enzyme into its active form. However, researchers followed two dead ends before arriving at the TB connection. The first was cholera.

It made sense that being a CF carrier might protect against a diarrheal disease, as the CFTR protein affects the composition of body fluids, and diarrhea is a massive loss of water from the digestive tract. Might CF carriers resist cholera, a water-borne infection responsible for outbreaks that rapidly kill thousands? The bacterium, *Vibrio cholerae*, under certain conditions produces a toxin that binds to normal CFTR ion channels in intestinal lining cells. This releases water in a diarrheal rush that can cause circulatory collapse, and death, in just hours or days. The diarrhea is colorfully described as "rice water," because of the white flecks of intestine in the stool. As *delta F508* carriers do not have the normal number of CFTR proteins, perhaps cholera toxin cannot bind enough cells to cause diarrhea. However, unlike the elegant sickle cell/malaria tale, geographical maps of cholera and CF do not align, and cholera arrived in Europe too recently—in 1831—to have had such a powerful effect in selecting *delta F508* alleles (Quinton 2007: S3).

The diarrhea idea persisted, and researchers followed a second dead end. Could CF carriers resist a different condition, perhaps typhoid fever, caused by the bacterium *Salmonella typhi* (Pier et al. 1998: 79)? Again, the bacterial toxin binds CFTR ion channels, but the maps of geographic distribution of the two diseases do not align. In fact, some populations riddled with typhoid fever, such as Indonesia, have no *delta F508* mutations at all.

Researchers went back to brainstorming why and how CF carriers were at an advantage. Balancing selection must satisfy three criteria:

1. a plausible molecular mechanism;
2. clinical correlation between the genotype and the infectious illness;
3. similar geographic distribution of the two diseases.

The last criterion rules out cholera and typhoid fever, but tuberculosis fulfills all three:

- *Molecular mechanism*: People with CF lack arylsulfatase B activity; carriers have half the normal amount. The TB bacterium,

Mycobacterium tuberculosis, requires the enzyme. Therefore, the bacteria can't survive in a human body without the enzyme.

- *Clinical correlation between genotype and infectious disease:* CF carriers have lower TB mortality rates, and people with CF rarely get TB.
- *Similar geographic distributions:* The longer TB has been in a region, the greater the proportion of CF alleles that are *delta F508.*

TB has a long enough legacy to have fueled the increased prevalence of the protective *delta F508* allele. The "white plague" TB, present for at least 15,000 years, became very deadly from 1600 to the late 1900s, killing 20–25 percent of the European population. Growing cities fueled the spread of TB. Today's geographical distribution of *delta F508* tracks with the infectious disease: where TB is most prevalent, so are CF carriers. TB did not reach India or Africa until the late 1800s, and here *delta F508* is rarer. The power of CF's protection was that TB, before the twentieth century and the discovery of antibiotics, was a deadly, systemic illness that killed people before they could reproduce. The infectious disease, therefore, would preferentially remove people *without* CF mutations from the population, leaving the legacy of cystic fibrosis today.

Considering modern disease prevalence statistics against a geographical backdrop is one type of evidence in population genetics. Another type of evidence consists of historical records. Eric Poolman and Alison Galvani, at the Yale University School of Medicine, consulted historical mortality records to evaluate the three proposed infectious diseases responsible for maintaining *delta F508* alleles: cholera, typhoid, and TB (Poolman and Galvani 2007: 91). They concluded that only TB can account for today's high prevalence of CF carriers in Europe, and that the pattern of the spread of the TB pandemic from 1600 to 1900 explains the drop off in modern CF incidence outward to the rest of the world. But the ability to treat TB removes the balancing selection, because people without CF mutations can survive. Poolman and Galvani developed a mathematical model that predicts that, if TB remains treatable, the incidence of CF will fall by about 0.1 percent per year for the next century. Antibiotic treatment has quelled the selective pressure—but the rise of multi-drug resistant TB, in the wake of the AIDS epidemic, could counter that. Another aspect of CF that may have also influenced today's concentration of mutant alleles in northern climes is that carriers make about half the normal amount of sweat. That would have been a problem in the tropics.

A third type of evidence in population genetics is rare—preserved ancient DNA. Such evidence indicates that *delta F508* existed long before the Renaissance. Analysis of the *delta F508* region of the CFTR gene from molar teeth removed from 32 skeletons, found in Austrian cemeteries dated to 544–255 BC, revealed the allele in three of them (Farrell et al. 2007).

Clues to the time and place of origin of different CF mutations lie within the DNA sequences elsewhere in the gene, and in surrounding regions on chromosome 7. This genetic background is termed a **haplotype**. All *delta F508* chromosomes have the same haplotype, which points to a single origin of that mutation, because a gene and its neighbors are usually transmitted together on their shared chromosome, like people linking arms and walking down a street. A modern population group where the *delta F508* haplotype is the most prevalent is likely an original population. Several investigations point to the Baluchis, an ethnic group from western Pakistan that emigrated to the United Arab Emirates, eventually reaching Europe (Saleheen and Frossard 2008: 157). Today, their descendants live in Iran, Afghanistan, and Pakistan. The *delta F508* haplotype exploded in European populations under the selective pressure of tuberculosis. Other CF mutations have other origins—such as ancient Phoenicia and among the Celtic tribes.

BIOETHICS

This chapter has explored the levels of human genetics through the example of the most common, "lethal" single-gene disease, cystic fibrosis. CF can also be used to illustrate the effects of genetic technology and our interference with nature at all of these levels—molecules and cells, individuals, families, and populations.

At the molecular level, testing for the *delta F508* mutation became possible just a few years after its discovery. Today, screening prospective parents, fetuses, and newborns for CF is routine in several nations, including the UK, US, France, Italy, Spain, Austria, Poland, Czech Republic, Australia, and New Zealand. Tests are tailored to include mutations that are most common in particular populations.

At the family level, people use genetic information differently. Two people who want to have children together and discover that they are carriers of CF mutations associated with severe disease have several choices. They might avoid having a sick child by using an assisted reproductive technology (Table 3.8). Or, they may terminate a pregnancy. A third alternative is to plan treatment, as Mikayla's parents did. In places where pregnant women are not tested for CF, newborns with the disease may be diagnosed following a blood spot test for abnormal trypsinogen, a pancreatic enzyme. A sweat test and DNA test confirm the diagnosis. Such early diagnosis leads to prevention, and potentially lifesaving treatment, of symptoms. Had Mikayla's intestinal perforation not been detected before her birth and her delivery moved up, she might not have survived.

At the population level, our interventions have already altered the evolutionary course of CF. Incidence of the disease in general, and of the severe *delta F508* allele in particular, is declining sharply in countries that test for CF mutations. In Italy, years of carrier screening have led to a sharp decrease in the number of children born with CF (Castellani et al. 2009: 2573; Liou and Rubenstein 2009: 2595). In the US, prenatal testing and newborn screening in the state of Massachusetts halved the number of double deltas born, and the number of babies born with CF fell by more than 20 percent (Hale et al. 2008: 973).

Table 3.8 Assisted reproductive technologies

Technology	Description
Intrauterine insemination	Donor sperm does not have a mutant CF allele.
Surrogate mother	A woman has her egg fertilized by the carrier father's sperm and carries the pregnancy, then gives the child to the couple.
Preimplantation genetic diagnosis	Carriers donate egg and sperm for *in vitro* fertilization. Fertilized ova divide to 8-cell stage, then one cell from each is tested for the CF genotype. An embryo that does not have the genotype is selected and transferred to the woman's uterus.

Although population testing may solve a perceived problem within a family, such as preventing the birth of a child who will be very ill and suffer, it ultimately has societal effects. Such unnatural selection removes many of the most severely affected individuals from the population—such as Mikayla. This may open up resources for the people who have the disease, but it also makes these individuals rarer—and therefore perhaps more likely to experience discrimination. At the same time, as the years pass and the trend to remove the most severe CF genotypes continues, the people available to participate in clinical trials of new treatments will no longer be the sickest. How will researchers tell whether a new drug is actually working, or just seems to because the participants are relatively healthy?

Eugenics is the intentional altering of the gene pool to improve society. Responses to genetic testing can change the gene pool too, but the goal is to prevent suffering. Genetic testing is not, then, eugenic in intent, but, perhaps, may ultimately have that effect. The bioethical concern is the nature and meaning of the terms "improve" and "suffer"—these are highly subjective. With the human genome sequence known and tests to learn our personal genotypes already widely available, let us hope that we can use that genetic information wisely.

WHAT WOULD YOU DO?

Several reports in medical journals over the past few years have clearly shown a decrease in the number of newborns with CF, as a direct result of detecting couples where both partners are carriers, before they conceive. The incidence is falling because couples use various tests and technologies to ensure that their offspring do not inherit the disease. However, prenatal testing for CF is not yet routine everywhere, and, in these places, newborn screening sometimes identifies CF by testing for telltale chemicals in a sample of blood taken from a baby's heel. Treatment can then begin immediately. Many dozens of diseases are screened for this way.

In early 2010, a news media outlet in the US took advantage of the fact that many people are unfamiliar with prenatal and newborn screening tests, using CF as an example. A distraught mother who discovered that her newborn was a carrier for CF bemoaned government intrusion on her privacy, even though she had signed an informed consent statement—in the excitement of the birth, she just didn't remember it.

Do you think that newborn testing for genetic diseases harms babies or families, or helps them? How would you correct the problem that the media pointed out, that new parents often do not recall, or understand, the details of the papers that they sign? How can newborn testing be beneficial? Do you see a conflict in the success of treatments for CF and the fact that people are choosing to end pregnancies when they receive a prenatal diagnosis?

SUGGESTED READING

This chapter used one disease to discuss the levels of genetics. *Human Genetics: Concepts and Applications*, 9th edition, by the author (McGraw-Hill Higher Education, 2010), presents this material in textbook format in the introductory chapter. *Alex: The Life of a Child*, by Frank DeFord (Viking Press, 1983) chronicles the declining health of the author's eight-year-old, who had CF. "My so-called lungs" is an audio diary by Laura Rothenberg on National Public Radio's *All Things Considered* program from 2003; she died of CF at age 22. However, most people with CF fare much better than either Alex or Laura. Learn about them, and the great strides in CF research, at the website for the Cystic Fibrosis Foundation (http://cff.org/).

GENES AND HEALTH

A human body is a complex, living machine. It is built of a few types of chemical element: carbon, hydrogen, oxygen, nitrogen, sulfur, and phosphorus. These elements combine into a few types of large molecule: nucleic acids, proteins, carbohydrates, and lipids. From these building blocks arise our 260 or so different types of cell, which assort and assemble into four types of tissue, which in turn interact to form our organs and organ systems. The nucleic acids DNA and RNA control the activities of our bodies through the production of proteins. Some proteins are like building materials, such as the stretchy connective tissue protein elastin or the scaly skin protein keratin. Very important are the enzymes, which are proteins that control the rates of the hundreds of chemical reactions that keep us alive and well.

When a gene is abnormal in a way that changes how its encoded protein functions, health may fail. The nature and location of symptoms depend upon which cells normally produce the affected protein. Mikayla Polski, discussed in Chapter 3, experiences breathing and digestive difficulties because her cystic fibrosis mutations alter the secretory activities of cells lining the passageways in her lungs and pancreas. Yet genes do not function in isolation, because a person's environment powerfully influences health. If Mikayla breathed smoky or polluted air, her respiratory symptoms would be far worse than they are in the safe environment that her parents have provided.

CAUSES OF DEATH—AN ORGANIZING FRAMEWORK

A discussion of genes and health can go in several directions. A popular book explores one health-related gene per chromosome—not terribly meaningful. A database called "Online Mendelian inheritance in man" describes every single known human gene, offering perhaps too much information. A textbook might distinguish single-gene disorders by their inheritance pattern—recessive or dominant, autosomal or sex-linked.

The health care community organizes diseases into medical specialties based on organ system. A gastroenterologist focuses on digestion, an endocrinologist on hormonal disorders, a dermatologist on skin. Medical specialists also define their turf by age, from pediatrics to adolescent medicine to internal medicine to gerontology. In contrast, the conditions that a medical geneticist encounters are more likely to span multiple organ systems than the problems a specialist confronts. Within any one medical specialty, single-gene disorders are rare.

Another way to classify diseases is by the degree of genetic influence (Figure 4.1). A Mendelian trait or disease is caused by a single gene, and a polygenic trait is caused by more than one gene. A trait or disease is multifactorial if it is caused by one or more genes plus environmental influences. One of the genes may have a large impact, and the others very little, or many genes can each contribute to a small degree to a trait. Very few traits are purely genetic. Often, researchers will learn about a medical condition by studying rare individuals with single-gene disorders, then apply what they learn to more common conditions. For example, a drug to treat emphysema might be tested on people who have a very rare, inherited enzyme deficiency that causes this lung disease, because the mechanism is more straightforward, and then be tested on people who have emphysema attributed to smoking.

Yet another way to look at health and disease is epidemiologically—the most common causes of death. This chapter combines this approach with the genetic view, and considers the three most common

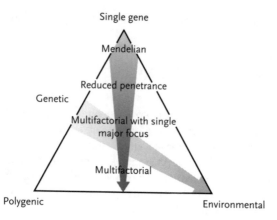

Figure 4.1 Genes/environment. Genes never function alone; the environment exerts many effects on gene expression. Penetrance refers to the fact that not everyone with a particular genotype expresses the associated phenotype. A multifactorial trait is determined by at least one gene plus environmental factors. A trait determined by more than one gene, with or without environmental input, is called polygenic

Source: Adapted from *Applied Genetics in Healthcare*, Figure 13.2, p. 225

classes of disease worldwide: cardiovascular disease (the heart, blood, and blood vessels), cancer, and infectious disease.

The most common causes of death reflect economics as well as biology. Although cardiovascular disease is the most common cause of death in all nations, the second most common cause of death in high-income nations, such as the US and the UK, is cancer, with infectious disease a distant third. In low-to-middle-income nations, deaths from infectious diseases are almost as common as those from cardiovascular diseases. These data can also be stratified. For example, of the more than 12 million children who die each year worldwide, 99 percent live in low-to-middle-income nations, and more than half of them had any of just five infectious diseases: respiratory infections, measles, diarrhea, malaria, and HIV/AIDS. These diseases are common in poor nations, but very rare in wealthy ones.

Each of the three classes of disease presents unique challenges to the body, and genes influence them all. The cardiovascular system is a pump and a network of pipes that distribute an oxygen-bearing fluid, blood. Cancer stems from an imbalance between cell division and cell death. Infectious diseases are the consequence of our sharing the planet with others. Genes orchestrate our immune defenses against the pathogens that use their genomes to inhabit our bodies.

The following sections consider selected examples of these three themes—cardiovascular disease, cancer, and infectious disease—at three levels: single genes, multifactorial influences, and a genome-wide view.

CARDIOVASCULAR DISEASE

Perhaps so many of us succumb to heart-related diseases because the cardiovascular system is vast and its regulation complex. Diseases result from mechanical failure of the heart or a disturbance of its electrical activity; deposition of fatty plaque on blood vessel linings or weakenings in their walls; or blood that is too thin or too prone to clot, or moves through the labyrinth of blood vessels under too much pressure. Hundreds of genes contribute to cardiovascular health, including at least 50 that maintain blood pressure. Single-gene diseases of the cardiovascular system are rare. The more common disorders typically reflect the actions, or failures, of several genes, each contributing a small degree of risk, coupled with environmental risk factors, such as a nutrient-poor diet, smoking, and lack of exercise.

Table 4.1 is a sampling of single-gene disorders of the heart, blood, and blood vessels. Most of them can be caused by mutations in any of several genes. Familial hypercholesterolemia (FH) is of particular significance because it led to development of a class of drugs, called statins, that help millions of people *without* inherited cardiovascular disease maintain healthy levels of blood serum cholesterol. FH, caused by a mutation in one gene, is extremely high blood serum cholesterol.

A RARE DISEASE

In the 1980s, two people with FH made headlines—Stormie Jones and Jim Fixx. Stormie was born in 1978 in Texas. The first sign of a problem was the yellowish bumps on the baby's buttocks, which Stormie's mother showed to the pediatrician at the three-month

Table 4.1 Single-gene disorders of the cardiovascular system

Disease	Defect
Blood clotting:	
Hemophilia A	Bleeding from deficient clotting factor VIII
Von Willebrand disease	Bruising from deficient plasma protein
Blood composition:	
Erythrocytosis	Increased number of red blood cells
Fanconi anemia	Decreased number of red blood cells
Blood vessels and circulation:	
Marfan syndrome	Absent or deficient fibrillin (a connective tissue protein) weakens wall of largest artery, which may burst
Sickle cell disease	Abnormal beta globin bends red blood cells into sickle shapes, blocking circulation
Heart structure and function:	
Cardiomyopathy	Abnormal cardiac muscle enlarges heart and impairs pumping
Long QT syndrome	Malfunctioning potassium channels in heart muscle disrupt heartbeat
Lipid profiles:	
Familial hypercholesterolemia	Too few LDL receptors cause very high serum LDL cholesterol
Hyperlipidemia	Elevated blood serum LDL, VLDL, and triglycerides
Hypertriglyceridemia	Elevated blood serum triglycerides
Lipoprotein lipase deficiency	Elevated blood serum LDL and triglycerides

check-up. The doctor assured the anxious new mother that nothing was wrong, and that the bumps would disappear. They didn't. By age five, Stormie had the odd, yellow deposits in the crooks of her elbows, knuckles, and knees, and between her toes. Finally, Dr. David Bilheimer, the medical director at Parkland Memorial Hospital in Dallas, recognized the bumps as pure cholesterol and found that Stormie's blood serum cholesterol was nine times the normal level. Listening through his stethoscope, Dr. Bilheimer heard the telltale sounds of dangerously blocked arteries. Stormie had blood vessels like those of an old person with heart disease, thanks to inheriting two mutations in a gene that gave her the severe form of FH. She was one in a million.

The little girl spent much of her time in hospital. By age six, Stormie had suffered two heart attacks and had severe chest pain. At age seven, triple bypass surgery replaced the arteries in her heart, and she received a new heart valve. Then she underwent the first heart and liver transplant ever performed—she needed the liver because it is the site of the body's own cholesterol production, and Stormie's made too much, and she needed the heart because her own was too weak to have survived a liver transplant. After spending several months in or near the Children's Hospital of Pittsburgh, where she had the transplant, Stormie was able to return home. She lived more normally, even returning to school, until she died at age 13 because her body rejected the heart.

At about the same time that Stormie Jones received her historic transplant, a runner credited with starting the jogging craze of the mid 1980s died of a heart attack. Jim Fixx's *The Complete Book of Running* had encouraged many people to give up cigarettes and fatty diets to start running, and the fact that the author had become an icon of healthful living made his sudden death in 1984, at age 52, shocking. Fixx had rarely seen a doctor, so little was known of his medical history, but he had been a two-pack-a-day smoker. When physicians learned that his father had died at age 43 of a heart attack and had suffered his first at age 35, they realized that Jim Fixx likely had the less severe form of FH, having inherited one mutation. The autopsy supported this idea: Fixx's coronary arteries were almost completely blocked with plaque. In contrast to the rarity of homozygotes such as Stormie Jones, one in 500 people is a heterozygote for FH, facing an elevated risk of a heart attack in early adulthood, unless they can control their cholesterol levels.

THE STATIN STORY

Cholesterol becomes a problem when it accumulates in the bloodstream, rather than being used inside cells. Our bodies require cholesterol, a lipid, to synthesize steroid hormones (such as estrogen, testosterone, and cortisol), bile, and vitamin D, and it is a major component of cell membranes. The liver manufactures cholesterol, and we get it in food. When cholesterol is so abundant that it overwhelms the receptors and cannot enter cells fast enough, it backs up and builds up as sticky plaque on the interior surfaces of the arteries. Blood must then squeeze through an ever-narrowing tunnel as it circulates, raising pressure as artery walls strain to contain the torrent. The plaque buildup and blockage is atherosclerosis, which is the primary cause of coronary artery disease.

Cholesterol is transported in the blood and lymph aboard lipoproteins, which are molecules consisting of a fat (lipid) attached to a protein. Bulk distinguishes different classes of lipoproteins. We have very-low-density lipoproteins (VLDL), low-density lipoproteins (LDL), and high-density lipoproteins (HDL). The genetic connection is that genes encode the protein portions of these lipoproteins and also the receptors on cell surfaces that bind them, taking cholesterol into the cells. The research that unraveled the steps of the transport and delivery of cholesterol to body cells led directly to development of the statins. These drugs block the enzyme required for the liver to make cholesterol, sharply lowering blood serum cholesterol.

In 1856, German pathologist Rudolph Virchow (1821–1902) was the first to realize that cholesterol buildup in artery walls causes atherosclerosis. A century later, studies began to corroborate his work at the molecular level, correlating elevated LDL to increased risk of developing coronary artery disease (Keys 1980). Cholesterol became the enemy, a target in the fight against heart disease, and the first cholesterol-lowering drugs soon appeared. They were a mixed bag, including thyroid hormone, estrogen, the vitamin niacin, the antibiotic neomycin, various plant derivatives, and a compound called cholestyramine. These drugs worked in different ways, and most had unpleasant side effects that limited their utility. But they did work, and investigating how they did so revealed a clue that would lead to development of the statins.

In 1959, researchers at the Max Planck Institute in Germany discovered that cholestyramine binds bile acids, excreting them in feces. The liver senses the diminishing supply of bile acids and starts to replace them. This requires cholesterol, and to synthesize it, liver cells require an enzyme called HMG-CoA reductase. As the bile acid level falls, the cell makes more of the enzyme (Tennet et al. 1960: 469). Animal studies showed that a cholesterol-free diet forces the liver to make its own cholesterol, and that it is HMG-CoA reductase that "tells" the liver to increase cholesterol production (Siperstein and Fagan 1966: 602). Would blocking the enzyme lower blood serum cholesterol, because the liver's contribution would be reduced or eliminated? By 1970, it was clear that this approach, coupled with a low-fat diet, should solve the high-cholesterol problem. The strategy sounded safe, because the substance that accumulates when the enzyme is blocked is easily excreted in the urine.

The statin story opened with converging lines of research at Tokyo Noko University in Japan, the University of Texas Health Science Center in Dallas, and the drug company Merck Research Laboratories. In 1971, Japanese microbiologists Akira Endo and Masao Kuroda were looking for natural inhibitors of HMG-CoA reductase. They were working with a type of fungus that makes a molecule which prevents another type of fungus from making certain lipids, including cholesterol. In this way, the first fungus outcompeted the second for its "turf." Such a "natural product" is often the source of a drug, or the inspiration for synthesizing one. Over two years, Endo and Kuroda screened 6,000 organisms for natural products that block cholesterol synthesis in cells. The first one they found, from a root rot fungus, wouldn't make a good drug, because its effects are irreversible. Complete cholesterol blockage would be deadly. Then, the researchers found a compound from another fungus, *Penicillium citrinum*, that inhibits the enzyme. They called it compactin at first, then changed it to mevastatin for mevalonic acid, the first compound that results from the enzyme's breakdown. By 1973, the researchers had deduced the three-dimensional structure of the mevastatin molecule.

In 1973, while Akira Endo was pursuing HMG-CoA blockage, Michael S. Brown and Joseph L. Goldstein at the University of Texas were examining cells from patients with the severe form of FH. The researchers discovered that the defect behind the disorder was in the cell membrane receptor for LDL, encoded by a gene on chromosome 5 (Brown and Goldstein 1976: 150). People with severe disease, such as Stormie Jones, were homozygous recessive. People who suffer heart attacks at early ages and have relatives who do too, like Jim Fixx, may be heterozygotes for this gene. Brown and Goldstein won the 1985 Nobel Prize in Physiology or Medicine for the discovery, because it shifted thinking about elevated cholesterol from a blood problem to an LDL receptor problem. "Brown and Goldstein's discoveries have lead to new principles for treatment, and prevention, of atherosclerosis," wrote the Nobel committee.

Discovery of the LDL receptor explained how cholesterol forms plaque on artery linings—both in FH and in people who eat too much fat. The LDL receptor is a protein that forms a coated indentation in the cell membrane, into which an LDL particle fits and is then taken inside the cell. Here, the trash disposal organelles, the lysosomes, degrade the protein part of LDL, releasing its cholesterol cargo. In liver cells, some of the liberated cholesterol blocks the activity of HMG-CoA reductase, essentially signaling "enough cholesterol came in with the fish and chips, there's no need to make any more." If all is well, cholesterol that is not taken into cells aboard lipoproteins and stays in the bloodstream binds to HDL, which carries it to the liver or out of the body. But, if the blood has too few HDL particles, or if the LDL particles cannot escort cholesterol inside cells for use sufficiently well, cholesterol builds up in the bloodstream. This is why LDL is referred to as the "bad" cholesterol, and HDL is the "good" cholesterol.

In FH, the shape of the LDL receptor distorts so that it cannot effectively bind LDL, and the level of LDL cholesterol in the bloodstream skyrockets (Figure 4.2). Other genes determine the number of LDL receptors, which explains why some people can eat a fatty diet but have healthy cholesterol profiles—they inherited many LDL receptors—and some people watch their diets very carefully yet have high cholesterol—they inherited too few LDL receptors. That is, low LDL and high HDL profiles promote cardiovascular health, but attaining this is naturally easier for some of us than others.

The researchers in Texas and Japan collaborated when they realized that they were looking at different aspects of the same problem. Cells from FH patients provided a way to test candidate cholesterol-lowering drugs that would be closer to the human condition than the animal models that Endo's group had been using in experiments. Mevastatin didn't work in mice and rats, whose biochemistries had ways to make cholesterol even if the targeted enzyme was blocked. The researchers had better luck with hens— they turned to chickens because of the high cholesterol levels in the eggs. On mevastatin, the cholesterol level in hens plunged by half, and that in dogs and monkeys by a third.

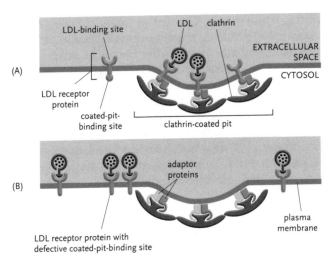

Figure 4.2 LDL receptor. LDL particles escort cholesterol inside cells for use and must bind to LDL receptors to do so

Source: *Molecular Biology of the Cell*, 5th edition, Figure 13–51, p. 791

By 1976, the Japanese researchers were sharing results with researchers at Merck, who confirmed their findings. Also that year, Endo's team gave mevastatin to a 17-year-old girl with severe FH who had a cholesterol level of 1,000 mg/dl. After two weeks of treatment, her cholesterol was down to 800. They and others tested mevastatin on several patients, with both severe and milder FH and other forms of high cholesterol, at different dosages. The drug worked well, even in people who had no symptoms.

Meanwhile, it was becoming clear that mevastatin was not the only way to block HMG-CoA reductase. Brown and Goldstein had discovered a similar natural product from a different strain of the same fungus that had yielded mevastatin. In 1979, researchers at Merck derived yet another HMG-CoA reductase inhibitor from "red yeast rice," which is actually a fungus, *Aspergillus terreus*. The Chinese had used red yeast rice for centuries to treat a variety of ills, including diarrhea, indigestion, spleen problems, and poor circulation. The Merck researchers named the compound monocolin K, which became known as mevinolin, and then lovastatin by the time the US Food and Drug Administration approved it as the first statin drug in

1987, marketing it as Mevacor. This was an analog of Endo's meva-statin, but better. By 1990, three statins were available, and, by 2006, the first, lovastatin, had gone generic after the patent expired.

Today, statins are the bestselling drugs worldwide. In the UK, they are the biggest item in the National Health Service's drug budget: 3 million citizens take them. In the US, 13 million people take statins. The world market exceeds 20 million and is poised to double, as studies suggest that even healthy people may benefit by controlling the level of cholesterol in the blood. They can thank the one-in-a-million people with a rare inherited disease for paving the way towards their daily cholesterol-lowering pill.

CANCER: THE CELL CYCLE OUT OF CONTROL

An organ, such as a liver or kidney, is shaped so distinctively that it is easily recognizable in different people, and at different ages. A kidney is clearly a kidney, whether in a baby, young adult, or elderly person. Organs maintain their shapes thanks to a highly regulated process, called the **cell cycle**, that determines when, and how often, a particular type of cell divides (Figure 4.3). A skin cell may divide every few days, yet a brain neuron, never. Cells have a built-in "clock" that tells them when to divide, in the form of chromosome tips, where copies of a six-base DNA sequence are shed with each division. The chromosomes shorten, like a fuse. When the chromosome tips, called **telomeres**, shrink to a certain point, division halts. Most human cells growing in laboratory culture divide up to about 50–60 times; similar limitations presumably operate in the body. The cell cycle also has built-in quality control measures. Some damage to DNA can be repaired, but, if the harm is too great, the cell dies by a process called apoptosis.

When control over how often cells divide is disturbed or lifted, cancer can result. Cancer cells have distinctive characteristics. They are less specialized than the cells from which they derive, and more likely to pile up on each other, forming lumps. Cancer cells invade surrounding tissue. They also secrete factors that coax nearby blood vessels to extend towards and then into them, which they use to enter the circulatory or lymphatic system. From here, cancer cells can spread, or metastasize, to distant parts of the body. Most importantly, cancer cells are "immortal," dividing continually.

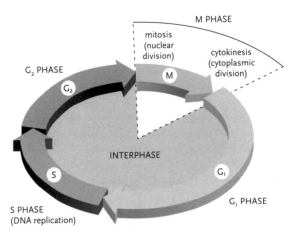

Figure 4.3 The cell cycle. Mitosis is when the cell divides. Interphase was once thought to be a time of cellular rest, but in fact it is a very active period, when the cell synthesizes many proteins

Source: *Molecular Biology of the Cell*, 5th edition, Figure 17–4, p. 1055

CANCER GENES

Genes are implicated in cancer in several ways. Only about 10 percent of cancers are inherited as a single-gene trait. When this happens, a person inherits a mutation that confers susceptibility to cancer in every cell, but the disease begins in a somatic cell, where a second mutation occurs in the same gene. Therefore, the cancer results from "two hits"—a susceptibility mutation in every cell, termed a germline mutation, and a second mutation in a somatic cell. (Recall that a somatic cell is any cell except a sperm or an egg; "somatic" means "body.") The other 90 percent of cancers also arise from two genetic "hits," but they are in the same somatic cell—that is, the mutations are only in the cancer cells. Mutations in somatic cells are usually caused by an environmental insult, such as exposure to radiation or to a cancer-causing chemical (a carcinogen).

Two types of gene cause cancer when they mutate (Figure 4.4). Proto-oncogenes are genes that normally control the cell cycle, but become cancer-causing **oncogenes** when they are overexpressed. Their products are called oncoproteins. In contrast, **tumor suppressor**

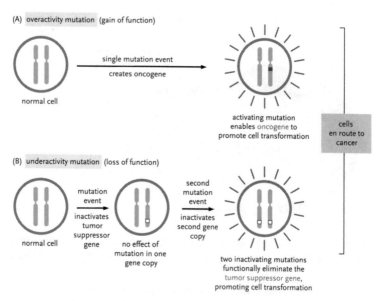

Figure 4.4 Oncogenes/tumor suppressors. Oncogenes cause cancer when they are overexpressed. Tumor suppressors cause cancer when they are deleted or are underexpressed. As a cancer progresses and spreads, many genes become involved

Source: *Molecular Biology of the Cell*, 5th edition, Figure 20–27, p. 1232

genes normally prevent cancer by holding down the number or rate of cell divisions, but, when they are inactivated or deleted, cancer results. Cancer may begin as a general destabilizing of the genome, which permits mutations to occur in proto-oncogenes and tumor suppressor genes.

Mutations in the same tumor suppressor genes or oncogenes may cause cancer in different body parts. In Li-Fraumeni syndrome, for example, family members who inherit a mutation in a gene called *p53* are at extremely high risk of developing any of several cancers. Half of them have at least one cancer by age 30, and 90 percent of them by age 70. People with the syndrome have the *p53* mutation in every cell, and cancers develop in cells where somatic mutations affect the same gene. Only about 100 families in the world have the syndrome. Although Li-Fraumeni syndrome, with its germline and somatic mutations in *p53*, is very rare, having two somatic mutations in the gene is fairly common. This may happen, for example, in a person who develops lung cancer after many years of exposure to cigarette smoke. Mutations in the *p53* gene also cause cancers of the thyroid, spleen, liver, esophagus, colon, bladder, and blood. The *p53* gene normally functions as a "cell cycle checkpoint," determining whether a cell with damaged DNA dies or is repaired. With a double mutation, p53 protein fails to condemn irretrievably damaged cells to death, and, instead, the cells become cancerous.

A SINGLE-GENE VIEW: BREAST CANCER

Breast cancer is often discussed as if it is one disease. Actually, mutations in more than a dozen genes can raise the risk of developing, or directly cause, breast cancer. The cancers often arise from mutations that disrupt DNA repair in a way that enables other mutations directly to cause the disease.

Like the p53 cancers, breast cancers more commonly arise from two somatic mutations, but we have learned about the disease process from the rarer, germline forms. About 5 percent of breast cancers are inherited, and about a fifth of these are caused by mutations in tumor suppressor genes called breast cancer predisposition genes 1 and 2 (*BRCA1* and *BRCA2*). They were discovered in the early 1990s in families in which several members developed cancer at young ages.

Like other rare, single-gene disorders, *BRCA1* breast cancer is much more common in certain populations. It is most prevalent among the Ashkenazi Jewish people, whose ancestors came from middle or eastern Europe. About 2 percent of this population has the mutation, compared with one in more than 800 in the general US population and about one in 1,000 in the UK. Even if a woman inherits a *BRCA1* mutation, she may never develop an associated cancer. For a woman with four Ashkenazi grandparents, the risk of developing cancer, if she inherits a *BRCA1* mutation, is 87 percent, but for a woman in the general US population, it is only 10 percent, and in the UK even less. These differences are due to influences of other genes and environmental factors.

A *BRCA* mutation is not just a concern for women. Males can develop breast cancer, but this is rare. For example, a man with a *BRCA2* germline mutation faces a 6 percent risk of developing breast cancer. This may not seem very high, but it is 100 times the risk for men in the general population. A man who has a *BRCA1* or *BRCA2* mutation, whether he develops breast cancer or not, still passes it to each of his offspring with a likelihood of 50 percent. The *BRCA* genes, when mutant, can cause other types of cancer. The same *BRCA1* mutation may cause breast cancer in one family member, ovarian cancer in another, and prostate cancer in yet another. *BRCA2* mutations can also cause cancers of the kidney, colon, skin, stomach, pancreas, prostate, or gallbladder.

Understanding DNA repair helps to explain how breast cancer can have so many genetic causes. Normal BRCA1, BRCA2, and p53 proteins form a complex in cell nuclei that peruses newly replicated DNA and identifies places where both sides of the double helix are broken, which could break the chromosome. DNA repair ensues. Mutations that affect any of these DNA repair proteins, plus others that interact with them, can cause breast cancer.

A MULTIFACTORIAL VIEW: COLON CANCER

In contrast to cancers that are triggered by mutations in a single gene, hereditary colon cancer is the culmination of a series of mutations that transform normal lining tissue into harmless (benign) growths that become precancerous and then cancerous. The multi-year process begins with a general destabilizing of the genome that activates certain oncogenes and disables certain tumor suppressor genes. It may begin with a mutation in a gene that maintains chromosomal structure during cell division, or in a DNA repair gene. A germline mutation in a DNA repair gene causes hereditary nonpolyposis colon cancer (HNPCC, or Lynch syndrome), which brings a lifetime colon cancer risk of 80 percent, with disease usually beginning before age 45.

Genome destabilization may also result from abnormal patterns of gene expression, rather than, or in addition to, mutations. The distinction is that mutations change the DNA base sequence; changes in gene expression alter which genes are turned on or off in cells lining the colon. Normally, small molecules called methyls silence parts of the genome by binding to certain DNA sequences, physically shielding them from being transcribed into RNA, the first step in protein synthesis. (A methyl is a carbon atom bonded to three hydrogen atoms, written CH_3.) In cancer, the control over where methyls sit on the DNA shifts. As a result, oncogenes may be activated, tumor suppressors silenced, or both. Altered methylation pattern is a type of **epigenetic** change, which means a change that affects gene expression, but does not actually alter the DNA base sequence.

Once the genome is destabilized, characteristic mutations ensue in a sequence, almost like a choreography of cancer (Figure 4.5). The first change is often in a gene called *APC*, which is a tumor suppressor and therefore normally blocks signals to the cell to divide. In cells where both copies of the gene are inactivated—either from a germline mutation plus a somatic mutation, or two somatic mutations—the smooth lining grows upward, forming a stalk-like structure called a polyp. Further mutations turn the benign polyp into a small adenoma, which is a precancerous growth. A gene that acts later in the process is *p53*, which enlarges the adenoma. Oncogenes are involved, too. Their activation affects enzymes called kinases, which control the signals entering cells. Some of these genetic changes can be detected in analysis of the DNA in stool samples. Another diagnostic test for colon cancer is a colonoscopy, in which a lit tube probes the colon for visible changes and growths.

CANCER GENOMES

Colon cancer is perhaps the best-studied cancer, its steps clearly recognized. By the time it has spread to distant organs, many mutations have occurred—on average, 15 mutations in "candidate" genes thought to directly promote the disease, and 61 in "passenger" genes thought to play a minor role. The number of involved genes is actually greater than this. When researchers scanned the genomes of colon cancer cells from 18,000 tumors, they found 848 genes that had somatic mutations! This means that many combinations of genetic changes cause and then fuel this common type of cancer. In addition, many of the oncogenes and tumor suppressor genes are not unique to a particular type of cancer. A "passenger" gene in colon cancer, for example, is a "candidate" gene in a glioma, which affects brain cells.

Cancer can happen in many ways, because gene functions overlap. One mutation is typically not sufficient to trigger disease, because

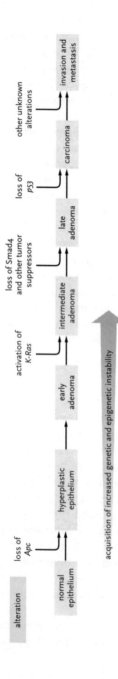

Figure 4-5 Colon cancer. Colon cancer progresses in a stepwise fashion. Oncogenes are turned on, and tumor suppressor genes are turned off

Source: *Molecular Biology of the Cell*, Figure 20–48, p. 1255

other, functioning genes can compensate, but, with enough mutations, cancer becomes inevitable. Smoking cigarettes for many years, for example, increases the risk of developing lung and other cancers by causing somatic mutations and, possibly, accelerating the mutation rate.

Understanding cancer genomes—the set of mutations that drive a particular cancer to spread—can reveal new drug targets. Researchers can consult the biology of the causative genes and determine where the relevant biochemical reactions and pathways converge. That point is the drug target. For example, if three oncogenes and five tumor suppressors all require enzyme X, then a drug that targets the enzyme will likely be more effective than drugs that target individual oncoproteins or tumor suppressor proteins. A very successful new class of cancer drugs, for example, targets tyrosine kinase, which is an enzyme that sends signals to divide. By plugging up and disabling the enzyme, the tyrosine kinase inhibitors have led to spectacular remissions in people with blood, stomach, and liver cancers.

INFECTIOUS DISEASE

Infectious diseases are much more commonplace than single-gene disorders, or even cardiovascular diseases or cancer. Nearly all of us have experienced an upper respiratory infection (a "cold"), influenza (the "flu"), or perhaps acne skin blemishes, a cold sore, Athlete's foot fungus, or a urinary tract infection. Common causes of death in many nations are the infectious diseases AIDS, tuberculosis, and malaria, caused by a virus, bacterium, and protozoan, respectively.

We suffer from infections because we live with many organisms that might take up residence in our bodies. These pathogens include single-celled organisms such as bacteria and amoebae; yeasts and other fungi; and "infectious particles" such as viruses, which are much simpler than cells.

THE IMMUNE SYSTEM

The immune system protects us against infections with a highly orchestrated series of general, as well as highly specific, responses to pathogens. The system consists of about two trillion cells, their protein products, and the organs that store the proteins. The genetic connection is that genes encode the two major types of immune system protein—**antibodies** and **cytokines**. The immune response is based on recognition of molecules called, generally, **antigens**. These are typically proteins jutting from the surfaces of cells or viruses, some linked to sugars, forming glycoproteins. Cell surface antigens form a topography of sorts, like the mountains and valleys of a landscape. Our own cells have antigen landscapes that the immune system perceives as "self" and does not attack; it reacts against foreign, or "non-self," antigen landscapes that are parts of pathogens.

The immune response creates a series of barriers that make it difficult for a pathogen to penetrate and colonize a human body. An infecting virus or bacterium must find a break in the skin or an exposed mucous membrane, and must overcome the likes of earwax, tears, forceful sneezes, and waving cilia, the tiny, hair-like extensions that sweep irritating particles and microbes up and out of the respiratory tract. Pathogens that survive these initial defenses then meet an acid bath in the stomach, and, if that doesn't get them, they may be forcefully expelled in vomit or washed out of the body in an explosion of diarrhea. In fact, much of what makes us sick is actually the response of the immune system, not an action on the part of the pathogen.

If a pathogen breaches or bypasses these physical barriers, it must face the reddening and swelling of inflammation. The changes of the inflammatory response are part of an initial, generalized, "innate" immune response that makes the human body less habitable. In inflammation, a rush of blood plasma brings in biochemicals that attack pathogens (Table 4.2), while diluting their toxins. White blood cells and giant cells called macrophages arrive from the circulation, engulf pathogens and draw them inside, where they are destroyed.

Innate immunity takes just minutes. Backing it up is the "adaptive" immune response, which is slower but highly specific. Fleets of white blood cells called **lymphocytes**—the T and B cells— carry out adaptive immunity. The T cells pump out cytokines and activate other cells to join the fight, including the B cells, which secrete antibody proteins that engulf and destroy specific pathogens. The precursors of T and B cells are in the bone marrow, and the T and B cells themselves patrol the entire body in body fluids. Of particular importance are T cells called "helpers," because they

Table 4.2 Proteins of the immune response

Antibodies	Built of Y-shaped subunits, antibodies bind non-self antigens, signaling other parts of the immune response
Collectins	Detect non-self antigens on certain viruses, bacteria, and yeasts
Complement	Plasma proteins that destroy viruses, rupture bacteria, and attract phagocytes
Cytokines:	
Colony stimulating factors	Signal bone marrow to manufacture lymphocytes
Interferons	Detect virally infected cells
Interleukins	Cause fever, which kills some pathogens
Tumor necrosis factor	Destroys bacterial toxins

activate B cells as well as secrete cytokines. Types of T cell are distinguished by their surface features, called "cluster of differentiation" antigens, such as CD4 and CD8. T cells are also classifed as "killers" or "suppressors."

An adaptive immune response begins when a macrophage—a giant, wandering cell—displays an antigen from a pathogen on its surface. This flag alerts a helper T cell. The T cell in turn binds and thereby activates a B cell that has surface receptors which can bind the pathogen antigen that the macrophage displays, like a key fitting into a lock. At the same time, the T cell secretes cytokines that signal the bound B cell rapidly to divide, producing two types of "daughter" cell: plasma cells that mass-produce antibodies that bind the displayed antigen, and "memory" B cells that snap into action should the pathogen appear again, protecting against future infection. Therefore, the antibody response is targeted and provides memory. It is also diverse. B cells can respond to nearly any provocation, because their antibody repertoire is so large. This is possible because DNA sequences that encode parts of antibodies can recombine in many different ways to create a great diversity of antibodies. The antibody response is a little like being able to create a vast number of outfits from a set number of shirts, skirts, leggings, jackets, and shoes.

VARIATION IN IMMUNE PROTECTION

Most of us do not think often about our immune systems. However, for people with inherited immune deficiencies, life is a daily fight against infection. Yet some lucky individuals inherit single-gene *protection* against infection. Just as studying rare, inherited forms of cardiovascular disease led to the invention of drugs to treat common conditions, studying rare individuals with single-gene disorders of the immune system has taught us about normal immunity. Inherited immune deficiencies affect all components of the immune response, from the initial generalized reactions, to innate immunity, to the specialized adaptive response.

Chronic granulomatous disease cripples the innate immune response by impairing the ability of phagocytes to destroy engulfed pathogens. Phagocytes cannot manufacture an enzyme that they require to produce chemicals, called reactive oxygen species, which kill certain pathogens that infect the skin, lungs, gums, and limb bones (*Salmonella, Klebsiella,* and *Burkholderia cepacia*), and fungi (*Candida* and *Aspergillus*). The missing enzyme is built of four parts, and so the disease can be inherited four different ways. Three of the genes are on the X chromosome, and so the condition is much more common in males. Before antibiotic drugs and vaccines were available, affected boys typically died in early childhood. Today, they can live well into adulthood, and can even be cured with a stem cell or bone marrow transplant that introduces a functional gene.

Twenty distinctive immune deficiencies target T and B cells. Eliminate B cells, and the antibody response fails. Eliminate T cells, and ultimately the entire adaptive immune response topples. Different types of immune deficiency make the body prone to specific infections.

A boy who inherits agammaglobulinemia, which is also X-linked, is typically healthy for the first month or two of life, because he still has antibodies received from his mother before he was born. Then, bacterial infections begin—ear infections, pneumonia, conjunctivitis ("pink eye"), sinusitis, and diarrhea. The boy is missing an enzyme that signals B cells to divide and mature into the antibody-spewing plasma cells. Agammaglobulinemia causes rampant bacterial infections because B cells can't make antibodies. In contrast, CD8 helper T cells fail in people with mutations in a gene called *DOCK8*. Viral infections take hold, especially warts, skin infections from certain pox viruses, and herpes simplex infections.

Severe combined immune deficiencies (SCID) eliminate *both* T and B cells. The original "bubble boy" was a young man with a form of the condition called SCID-X1 named David Vetter, who was born in 1971 in the US. Owing to a mutation on his X chromosome, he lacked a thymus gland. As a result, he could not make T cells, and so his B cells lay dormant. David, thoroughly without immunity, survived by living in a large plastic bubble. After he emerged to receive a bone marrow transplant from his sister, he developed lymphoma (a B cell cancer) from a virus transmitted in the bone marrow and died at age 13. Children with this and other forms of SCID have been successfully treated with gene therapy. Chapter 6 tells their stories.

Not all mutations are bad, despite the depiction of mutants in popular fiction. Some lucky individuals inherit variants in immune system genes that enable them to resist certain infections which make most people very sick. This is the case for infection by the human immunodeficiency virus (HIV), which causes acquired immune deficiency syndrome (AIDS). Since very early in the AIDS epidemic, researchers have paid special attention to people who are repeatedly exposed to HIV, but do not become infected. Studying them helped to reveal the steps of HIV infection.

HIV spreads through body fluids. The virus systematically attacks and disables the human immune system, demolishing protection against other infections and cancer. HIV affects more than 200 types of human protein.

HIV infection usually begins as the virus binds to two linked receptor proteins, CD4 and CCR5, on the surface of macrophages. In response, the cell engulfs the viral particle, taking it inside. The virus then commandeers the protein synthetic machinery, forcing the T cell into becoming a factory to mass-produce the virus, rather than providing innate immunity. Soon, the cell bursts, releasing viruses. Once macrophages have been subverted to boost the number of viruses, HIV next latches onto CD4 helper T cells, again binding and entering via the CD4 and CCR5 co-receptors. Unfortunately for us, these helper T cells are the linchpin of the immune system. Because they normally release cytokines, as well as activate the antibody defense, once their numbers start to fall, immunity severely falters. Bacterial diseases usually appear first, then viral infections and certain cancers. Symptoms start when the million or so new T cells that the bone marrow produces each day cannot keep pace with the number of new HIV particles exploding from infected T cells. The different drugs that are combined to treat HIV/AIDS attack the viruses at several key points as they enter and take over T cells. The sooner treatment begins, the longer the person is likely to live, and the less likely he or she is to transmit the infection.

People resist HIV infection in various ways. The first mechanism of HIV resistance identified was a mutation in the gene that encodes the CCR5 receptor that HIV must bind in order to enter cells. People who are missing 32 bases in both copies of the *CCR5* gene cannot become infected, simply because the virus cannot enter their cells, like a house without a door. In fact, a man who had AIDS and leukemia received a stem cell transplant to treat the cancer from a donor who was homozygous for the *CCR5* mutation—HIV vanished from his body (Hutter 2009: 692). That led to a new treatment approach—introducing the mutation.

About 1 percent of Caucasians are *CCR5* double mutants, and therefore HIV cannot infect them. The mutation is much less prevalent, or absent, in other populations. How did this happen? The frequency of the *CCR5* mutation among fair-skinned people is far too high to have arisen by natural selection (survival advantage) as a response to HIV infection, because natural selection takes a very long time, and HIV is fairly new in human populations. An alternate explanation might be that the protective double *CCR5* mutant genotype conferred a survival advantage in times past to people facing a *different* pathogen that used the same receptor to enter cells. Indeed, the *CCR5* mutation is most prevalent in parts of Europe that lost the most people during plagues of viral hemorrhagic fevers that swept through populations during the Middle Ages. Consider Eyam, a town in the UK where half the people died of such a plague in 1665. The *CCR5* mutation is greatly overrepresented among its population's descendants. After the Middle Ages, these viral fevers resurfaced in parts of Sweden, Denmark, Hungary, Poland, and Russia that, today, have a higher prevalence of the *CCR5* mutation. (The bacterium that caused bubonic plague does *not* bind the CCR5 receptor.)

Several studies have identified other inherited quirks of immunity that protect people against HIV infection (Cohen 2009: 1476).

A two-decade-long study of healthy female sex workers in Nairobi, repeatedly exposed to HIV on the job, revealed that they have unusual CD4 T cells that protect other T cells. Intravenous drug users in Vietnam, who never contracted HIV, despite repeated exposure, have extra natural killer cells, which are part of the innate immune response. Healthy men in Stockholm, repeatedly exposed to HIV through oral sex with infected male partners, have anti-HIV antibodies in their saliva. More mysterious is how 5 percent of people with hemophilia who received HIV in blood transfusions in the early 1980s avoided infection.

Over time, these various protections against HIV infection will likely increase in many populations, tempering the AIDS epidemic. More immediately, discovering how these inherited quirks protect people can suggest new ways to fight HIV infection.

A GENOME-WIDE VIEW OF INFECTIOUS DISEASE—LEPROSY AND INFLUENZA

During an infection, two genomes interact—that of the person and that of the pathogen. Susceptibility to leprosy (a bacterial infection) and influenza (a viral infection) illustrates the roles of the human and pathogen genomes.

The bacterium *Mycobacterium leprae* causes leprosy, which affects the skin and the peripheral nerves. In a mild form, small, pale, numb areas dot the skin, but, in a severe form, larger skin lesions full of bacteria lead to disfigurement. The advent of a triple antibiotic "cocktail" in the 1980s brought the number of global leprosy cases down from 5.2 million in 1985 to about 215,000 in 2009 (Stone 2010: 939).

About 95 percent of people are naturally immune to leprosy. Many of the susceptible 5 percent have affected family members and, in those cases, increased prevalence of blood relatives who have had children together. These observations suggest an inherited risk to contracting leprosy. The disease is particularly difficult to study, because the bacteria do not grow in the laboratory, and only infect humans and armadillos. For these reasons, and because the bacteria are very uniform genetically, researchers from the Genome Institute of Singapore and 26 institutes in China looked to the human genome to explain leprosy susceptibility, rather than the hard-to-study bacteria.

The researchers used an approach called a **genome-wide association study** (GWAS) that looks for single-base positions in the genome where people with a particular trait or illness share a rare base. Such a site is called a **single nucleotide polymorphism**, or SNP (pronounced "snip"). For example, if the 117th base from one end of chromosome 17 is a T in 96 percent of a population, then the 4 percent who have a C instead have a SNP there. People can be distinguished by large sets of SNPs. In the leprosy genome study, researchers scanned 2.4 million SNPs among more than 10,000 people, including those with either form of leprosy and healthy controls. Analysis of the SNPs that the people with leprosy share led the researchers to seven genes whose variants contribute to risk of contracting leprosy (Zhang et al. 2009: 2609). Some of the genes cause inflammation, and the others enable T cells to recognize an antigen that is unique to the cell walls of the leprosy bacterium. The findings may have broad applications. "Though leprosy is not common, the discoveries have significant ramifications for chronic infectious disorders and for host–pathogen interactions in other more prevalent mycobacterial diseases, such as tuberculosis," says Edison Liu, MD, executive director of the Genome Institute of Singapore.

Much more common than leprosy is influenza. Every year, many of us queue up to get the flu jab. A vaccine, such as that given to protect against influenza, is a disabled form of a pathogen, or part of it, that "fools" the immune system into reacting as if infection were taking place, signaling B cells to produce antibodies. The vaccine jumpstarts immune memory, so that if the person encounters the actual pathogen, reaction to squelch infection is so fast that the person is not even aware of the threat.

Vaccines date back to eleventh-century China. Healthy people rubbed smallpox crusts into their abraded skin, to induce a mild case that would protect against a more severe one. In 1796, British doctor Edward Jenner improved the method by using lesions from cowpox, a milder illness, after noticing that milkmaids who developed cowpox did not contract smallpox. Jenner's vaccine eventually rid the world of smallpox.

Immunization against influenza presents a different challenge from smallpox, because the influenza virus is very changeable, so much so that vaccines must be reinvented yearly. Because genetics is responsible for the rapid evolution of flu viruses, we can apply knowledge of the influenza genome to predict epidemics. Influenza viruses infect only a few species, including birds, pigs, ferrets, and us.

Three families of influenza virus are recognized—A, B, and C. Only A and B infect humans, with A causing most cases of the flu. Like many viruses, influenza A has a streamlined structure, compared with a cell. It consists of about 13,500 RNA bases in a protein coat. The RNA encodes just 11 types of protein. Human antibodies recognize and bind two types of glycoprotein on the viral surface. One type of surface glycoprotein, hemagglutinin (H), attaches to molecules of sialic acid on our cell surfaces. The other type, neuraminidase (N), releases new viral particles from infected cells. Flu viruses can have any of 16 types of H and any of 9 types of N. The most common combinations of influenza viruses that infect humans are "H3N2" and "H1N1."

Influenza viruses change at two levels. The smaller change, a single mutation, is called "drift." The mutation rate is high, because the virus replicates its genetic material frequently, and the enzyme that it uses to do so is error-prone—and the errors persist. A greater level of change, called a "shift," happens when two different flu viruses swap parts, creating a novel virus. The site of a shift is a pig's throat, where a bird (avian) flu virus might recombine RNA with a human flu virus. New flu virus variants arise each year in parts of Asia, where humans, birds, and pigs live in close proximity. A shift can generate a virus so unusual that the human immune system might not have antibodies to respond to it.

The World Health Organization convenes a global surveillance program twice yearly to identify the strains of flu in different geographical regions. Participants use the information to predict the predominant strains for the next season. The vaccine "recipe" they suggest is typically a mix of H1N1, H3N2, the rarer influenza B, and whatever else is circulating. But, sometimes, Mother Nature springs a surprise. That's what happened in April 2009.

The next season's flu vaccine was already in production when public health officials discovered a new, and very atypical, H1N1 strain in Asia and estimated that it had probably been in the human population since January. The new virus was an unusual mix of RNA sequences—part human H1N1, part Eurasian swine flu from 1979, part human H3N2, and, most alarming, part north American swine flu, frighteningly similar to the H1N1 virus that killed 50 million people in 1918. The genetic closeness to the 1918 virus is why a new, second flu vaccine was rushed into production and the world warned. Fortunately, "2009 H1N1" turned out to be highly transmissible, but usually caused mild disease. Efforts to keep ahead of flu virus evolution in the future will rely increasingly on information in the viral genome sequence, because this goes beyond the surface features traditionally used to distinguish viral subtypes.

Until very recently, medical genetics dealt with the rarest of the rare, single-gene disorders. Most physicians knew little about these conditions, and getting a diagnosis would often take years. Today, not only is genetics finally receiving attention in medical training, but health care consumers are much more aware of their family histories, even taking genetic tests on their own. We now know that even the common ills of humankind, which are not single-gene disorders, nevertheless unfold as a change in gene expression that parallels the course of the illness. In short, genes affect all aspects of our health.

WHAT WOULD YOU DO?

Fast, affordable sequencing of anyone's genome will soon be possible. But how useful will the information be?

Knowing your genome sequence is a little like having a list of every component of an automobile. If your car breaks down, the circumstance of the problem is probably more important than having the parts list. So it is for the human genome and what it can, and cannot, reveal about health. Consider the three types of illness discussed in this chapter. To what degree do errant genes cause them?

Cardiovascular disease might be completely inborn, such as little Stormie Jones and her sky-high cholesterol. However, much more often, it reflects a lifetime of unhealthy lifestyle choices, such as a poor diet and insufficient exercise. Lifestyle habits can also affect cancer risk, but many people who follow all of the rules still get cancer. Still, inheriting a cancer is never absolute. Women who face the possible consequences of inheriting a BRCA1 or BRCA2 mutation wrestle with the "incomplete penetrance" of their genotypes. Even a woman with such a mutation in the highest risk group, Ashkenazi Jewish people, faces an 87 percent lifetime risk of developing an associated cancer—not 100 percent. For most people, the risk is in fact much lower. When is the cancer risk high enough to justify having one's breasts and ovaries removed to prevent the disease?

We have better control over infectious diseases than we do cardiovascular diseases or cancer, largely because of the success of vaccines. Infectious diseases are treated with antibiotic or antiviral drugs, or supportive care until the immune system can fix the problem—not typically by consulting a genome sequence.

Human genome information should be placed into a practical perspective. Given the current state of medicine, how valuable would knowing a patient's genome sequence be to a physician in treating a cardiovascular disease, cancer, or infection? What are possible psychological problems that might arise from a genome sequence providing too much information? And finally, at the level of sociology, how should people be prioritized for having their genomes sequenced?

SUGGESTED READING

Reports appear daily of discoveries of gene variants and gene expression patterns that cause or contribute to a great variety of human ills. The following organizations offer up-to-date and reliable information.

CARDIOVASCULAR DISEASE

The American Heart Association (www.americanheart.org): Information under "diseases and conditions" and "healthy lifestyles" analyzes the environmental component to common disorders of the heart, blood, and blood vessels.

CANCER

Susan G. Komen Breast Cancer Foundation (ww5.komen.org/)

National Cancer Institute's Cancer Information Service (http://cis.nci.nih.gov/)

INFECTIOUS DISEASE

The National Health Service (www.nhs.uk. Influenza information)

The Foundation for AIDS Research (www.amfar.org/): This organization helps developing countries to promote HIV research, prevention, treatment, and education.

Centers for Disease Control and Prevention (www.cdc.gov/h1n1flu): Daily updates on the spread and course of this unusual flu.

J. Craig Venter Institute (www.jcvi.org/cms/research/groups/infectious-disease/): Sequences of pathogen genomes.

GENERAL

Sanger Institute (www.sanger.ac.uk/about/history/achievements.html): How 50 research organizations in the UK are discovering the genetic underpinnings of common diseases.

Genetic Alliance (www.doesitruninthefamily.org): A useful tool to see what's in your family.

World Health Organization:

- www.who.int/topics/cardiovascular_diseases/en/
- www.who.int/topics/cancer/en/
- www.who.int/csr/disease/swineflu/en/

5

GENETIC TESTING

The 78-year-old patient awakens. She has just undergone a diagnostic procedure, following admission to hospital with symptoms of a heart attack. She looks up into a kindly face. The doctor explains that the test found severely blocked arteries. The patient and her anxious family can now weigh the options: bypass surgery to replace the clogged blood vessels, insertion of a metal stent device to open them, or drug therapy. Fearful of the procedures, the woman chooses to manage her atherosclerosis—the buildup of plaque inside her arteries that caused the heart attack—with medications. The drugs cannot reverse the damage, but if she stops smoking and begins a healthier diet and exercise plan, the disease's course may slow.

The medical team knows which drugs are most likely to work and the least likely to have adverse effects for this particular patient, thanks to genetic tests. When the woman arrived at hospital, a technician took a small tube of her blood and placed it in a disposable cartridge about the size of a mobile phone. The cartridge contained reagents equivalent to a molecular diagnostic laboratory. The device performed standard tests, such as measuring levels of a muscle protein that, when elevated, indicates heart damage. The device also separated out the white blood cells in the sample and extracted DNA from them. It then identified genotypes that indicate sensitivity to various drugs that the woman would take for the rest of her life to maintain the functioning of her heart and blood vessels.

The technology to detect rapidly the genetic contributions to common disorders, such as heart attacks, is relatively new. Physicians are beginning to use it along with more traditional information, such as a family history, to diagnose, treat, and monitor response to treatment and disease recurrence. This type of testing for multi-factorial disorders—those that reflect genetic as well as environmental influences—is also new, because the first genetic tests, in the 1950s, detected rare, single-gene diseases. These were easier to understand. Multifactorial traits are also called "complex" traits, for good reason.

Then and now, genetic testing differs from other medical tests in several ways. Because nearly all of our cells contain our complete genetic instructions, any cell type can hold diagnostic clues for a disease that affects any part of the body. Because our complete genetic instructions are with us for life, mutations detected today can predict disease tomorrow. Finally, our personal genetic information applies to our relatives, with predictable frequencies for particular individuals.

In the future, genetic testing will be a crucial part of preventive health care and a key part of personalized medicine—as the heart attack patient learned. Her genetic tests were performed in a medical setting. Many DNA-based tests are available on the Internet and do not involve a health care practitioner. Therefore, it is helpful to understand the history of genetic testing.

THE FIRST GENERATION OF GENETIC TESTS: PROTEINS

The earliest genetic tests did not analyze DNA, but examined proteins, the products of genes. These tests inferred that particular enzymes were missing from buildups of the biochemicals they would otherwise break down, or they detected proteins with altered shapes that impair their function. The first three single-gene disorders to be screened for at the population level were phenylketonuria (PKU), Tay–Sachs disease, and sickle cell disease.

PKU—AN EARLY SUCCESS STORY

Pearl S. Buck (1892–1973) was a prize-winning novelist from the United States who spent most of her life living in, and writing about, China. She is best known for her book *The Good Earth*. Less familiar is her 1950 book *The Child Who Never Grew,* about her daughter Caroline, born in 1921. Caroline had a very limited attention span, and her eyes were oddly blank and unresponsive. She made purposeless movements. By the time the Bucks brought their daughter to the US for evaluation when she was three years old, she was clearly mentally retarded, and worsening.

Caroline Buck had PKU, a classic "inborn error of metabolism." Lack of an enzyme (phenylalanine hydroxylase) causes buildup of its substrate (the amino acid phenylalanine), as well as accumulation of an abnormal breakdown product (phenylpyruvate) (Figure 5.1). The fact that the accumulated substance is a normal component of dietary protein presents an opportunity to treat the condition with dietary restriction. This is why the ability to screen for the disorder at birth was a medical milestone.

For PKU, as for cystic fibrosis (discussed in Chapter 3), incidence and mutation types vary among populations (Table 5.1). Also like CF, PKU may be common in parts of Europe because carriers have enjoyed a survival advantage and, therefore, increased in the population. PKU carriers have elevated blood phenylalanine levels. It is not high enough to cause PKU symptoms, but sufficient to inactivate a poison found in rotting potatoes that is deadly to a fetus. In the great famines that swept Ireland and Scotland, might PKU carriers have better survived by eating moldy potatoes that did not make them sick, nor poisoned their unborn child? According to this hypothesis, with time, the incidence of this protective mutation

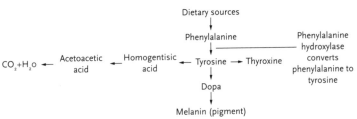

Figure 5.1 PKU pathway. In the inborn error of metabolism PKU, the amino acid phenylalanine cannot be broken down. Eating very little phenylalanine-containing protein from birth can prevent the mental retardation that the condition causes

Source: Adapted from *Genetics for Healthcare Professionals,* Figure 2, p. 122

would have increased and spread eastward as the Vikings took wives and slaves from the Celtic lands.

Caroline Buck was in her teens when PKU began to be understood. In 1934, the parents of two young children with "feeblemindedness," in Norway, noticed a musty odor to their urine. Six-year-old Liv walked strangely and spoke few words. Her four-year-old brother Dag seemed stalled in infancy, unable to walk, talk, or feed himself, and still wore nappies. The parents, Harry and Borgny Egeland, were perplexed, as both children had seemed healthy at birth. Harry told a colleague about his children's problems, who mentioned them to a physician and biochemist named Asbjörn Fölling. Intrigued, Fölling analyzed the urine in a makeshift biochemistry lab in an attic at University Hospital in Oslo. First, he ruled out infection. Then, he added ferric chloride to the urine to rule out diabetes, which would turn urine purple instead of the reddish-brown that normal urine turns. Liv's and Dag's urine turned green! Astonished, Fölling asked the parents not to give the children any medications for a week, and then he tested the urine again. It was still green! Using standard organic chemistry techniques, Fölling identified the substance responsible for the green color, and likely the odor, as phenylpyruvic acid. Could it cause mental retardation? He wondered where else he could find the anomaly.

Fölling next tested urine from 430 people institutionalized for mental retardation and found phenylpyruvic acid in eight, including two pairs of siblings. He named the condition *imbecillitas phenylpyruvica*. The affected individuals shared other character-

Table 5.1 Incidence of PKU in selected populations

Population	Incidence
Global	1/15,000
China	1/16,000
Finland	<1/100,000
Ireland	1/4,500
Japan	1/119,000
Norway	1/13,000
Sweden	1/30,000
Turkey	1/2,600
US Caucasian	1/10,000

istics—fair skin, rashes, stooped posture and spastic gait, and broad shoulders. Fölling published his findings on these ten patients (Fölling 1934: 169; 2008: 4), and, shortly afterwards, another researcher in the UK described three other families. The autosomal recessive mode of inheritance was clear: both sexes were affected, and the parents were healthy. The clincher was the fact that, for three sets of parents, husband and wife were blood relatives. They were passing on a recessive mutation inherited from shared ancestors.

Next, Fölling invented a better way to detect phenylpyruvic acid in urine, which became known as the "wet diaper test." It used bacteria that produce a colored compound when exposed to phenylpyruvic acid. The test could spot PKU, but not until the nappy-wetter was six weeks of age.

In 1937, George Jervis, the first director of the Institute for Basic Research in Developmental Disabilities, in Staten Island, New York, renamed Fölling's disease *phenylketonuria*, which means "phenylalanine in the urine." Jervis worked with 50 patients at Letchworth Village State School in Thiells, New York, adding 135 more patients from other state institutions (Jervis 1937: 944). He also noted that all children of women with PKU had small heads, mental retardation, low birth weight, and heart abnormalities. By 1952, biochemists filled in some of the blanks about how PKU arose (Undenfriend and Cooper 1952: 503). When phenylalanine can't be broken down, its breakdown product, the amino acid tyrosine, becomes depleted. As tyrosine is a precursor of melanin, the pigment molecule, people with PKU have fair skin and hair. The biochemical block affects other metabolites, causing hyperactivity, autistic-like behavior, and seizures.

In 1953, German physician Horst Bickel suggested that a low-protein diet might prevent mental retardation by decreasing the phenylalanine level (Bickel et al. 1953: 812). One of the first patients treated was a two-year-old at Birmingham Children's Hospital, in the UK, who could not walk or talk, cried piteously, and banged her head (Kayt Marra, personal communication and Ph.D. thesis 2010). Fed a formula very low in phenylalanine, but with a tiny amount of whole milk, eating only a few fruits, vegetables, and gluten-free grains, the girl started to crawl, haul herself up, and try to walk. Her head-banging stopped, and her hair darkened. When the diet accidentally lapsed, symptoms returned.

By 1958, the first commercial formula for PKU became available. However, a problem of timing emerged. The diet had to start very soon after birth to prevent mental retardation, yet the wet diaper test let too much time elapse, not working until six weeks of age. A blood test, called the Guthrie test, would prove much more accurate.

In 1957, Robert Guthrie was a microbiologist at the State University of New York at Buffalo. He and his wife had a son with mental retardation and a niece who had been diagnosed with PKU at 15 months, using the wet diaper test—much too late to prevent symptoms. At a meeting of the New York State Association for Retarded Children, Guthrie chatted with a physician working with people who had PKU, and had an idea. He had been investigating biomarkers in cancer patients' blood—biochemicals that would observably change a bacterial culture and so could be used to diagnose or monitor the cancer. Guthrie redirected his cancer project to detect elevated phenylalanine in the blood of newborns. The test used a strain of a common bacterium, *Bacillus subtilis*, that could grow only in the presence of phenylalanine, phenylpyruvic acid, and a related acid. If a newborn's blood had these substances, the bacteria would grow—and the child had PKU.

In 1961, Guthrie described his test at a New York State Association for Retarded Children conference in Jamestown, NY. Soon, two hospitals began sending him filter papers with blood drops from the heels of newborns. Then Guthrie ran his test on 3,118 institutionalized patients with mental retardation, finding 21 with PKU (Guthrie and Susi 1963: 338). A larger study validated the test in 400,000 newborns (Guthrie and Whitney 1964). At about this time, photographs from affected families began to show clearly disabled older siblings and younger ones who had obviously benefited from the diet. But it wasn't easy. The formula and drink, carefully manufactured to avoid phenylalanine, were rather unpalatable, and getting young children to stick to them without wailing at mealtime was an issue.

The Guthrie test caught on, fast. Circumstances helped, too, for President John F. Kennedy's mentally disabled sister, Rosemary, had drawn attention to the problem. Rosemary did not actually have mental retardation. At the age of 23, in 1941, a prefrontal lobotomy, performed supposedly to treat her mood swings, instead damaged her brain. Still, the Kennedy family has always championed treatment of those with mental deficiencies, in honor of Rosemary. The first mass newborn screening program for PKU in the US was in the Kennedy's home state, Massachusetts, in 1963. It spread quickly, and today all states require it. Newborn screening, teamed with treatment, was an economic success. In the early 1960s, the cost of screening all newborns in the US and treating those with PKU was about $3.3 million per year; the cost of not screening and then institutionalizing people with PKU was $189 million per year.

In 1983, Savio Woo and colleagues at the Baylor College of Medicine discovered the *PAH* gene, which encodes phenylalanine hydroxylase, the enzyme deficient in PKU (Woo et al. 1983: 151). Since then, researchers have identified more than 450 mutations in the *PAH* gene, associating different severities of PKU with different allele combinations.

Continuing investigation of the effects of the diet led to important discoveries. Unlike the initial protocol of following the diet only through childhood, and then only through adolescence, the regimen now is to do so for life, with careful selection of only very low phenylalanine foods and prepared drinks considered a "medical food." The PKU community learned the importance of maintaining the diet during pregnancy, when women with PKU who had stopped the diet in childhood gave birth to children with severe mental retardation, and even children who had not inherited PKU were affected. So much phenylalanine had accumulated in the mothers' systems, even though they were too old for their brains to be noticeably affected, that any offspring were essentially poisoned in the womb.

Despite the difficulty of following the diet, PKU newborn screening serves as a model for other genetic screening programs for several reasons:

- The test is easy, cheap and accurate.
- Consequences of *not* screening are costly and severe.
- Treatment is available.
- Early detection makes a difference in outcome.
- The test does not alarm parents, nor harm children.
- Genetic testing can supplement protein analysis to predict severity.
- Everyone is tested, not just specific population groups.

The Guthrie test for PKU developed in the 1960s is still performed in many nations on every newborn. The 1970s brought tests for other single-gene disorders. The most publicized programs differed from PKU newborn screening in one key way: targeting specific populations. Promoting a genetic test for a particular group of people, no matter how biologically sound, can open a door towards discrimination. Two genetic testing programs of the 1970s were especially telling. One was a success, and the other a disaster. The difference wasn't in the science, but in the way the tests were delivered.

TAY–SACHS DISEASE—THE FIRST "JEWISH GENETIC DISEASE"

In the spring of 1973, I took a carrier test for Tay–Sachs disease at my university on New York's Long Island. Tay–Sachs is a devastating degeneration of the nervous system that begins in infancy and is invariably fatal within a few years. Tay–Sachs screening in the US during my university days was part of a public education program to test students of Ashkenazi Jewish background, whose ancestors came from central or eastern Europe. This devastating inherited illness was once 100 times more common among the Ashkenazim than most other groups, and carriers are ten times more prevalent.

Tay–Sachs disease is autosomal recessive. I was planning to marry another Ashkenazim, so I knew that, if the test found I was *not* one of the one in thirty Ashkenazim who is a carrier, then he need not be tested—according to the laws of inheritance, we couldn't have a child with Tay–Sachs. The test detected the half-normal concentration of an enzyme, hexosaminidase A, in carriers, who are symptomless. Each child of a carrier couple faces a one in four chance of inheriting the disease. Fortunately, my blood had the normal amount of the enzyme.

Thanks to worldwide population screening, fewer than twenty infants are born a year with Tay–Sachs disease, and most or all of them are *not* Ashkenazim. The disease is so severe, and untreatable, that couples who learn that they are both carriers nearly always choose not to have children together, use egg or sperm from a third person, adopt children, or have prenatal diagnosis and end affected pregnancies. Before genetic testing, about 1 in 3,600 Ashkenazi newborns had the condition.

A baby with Tay–Sachs disease appears normal for the first few weeks. However, deep in his or her cells, in the tiny sacs called lysosomes where powerful digestive juices degrade debris, the enzyme deficiency causes a buildup of fatty material. Nerve cells in the brain and spinal cord slowly drown in fat, and the body weakens. By six months, the child stops trying to crawl or stand. Just sitting becomes difficult, as nerve messages to muscles grow silent. The child no longer pays attention when read to, and the eyes move strangely. Sudden sounds provoke an odd startle, perhaps even a seizure. At the backs of the child's eyes, "cherry red spots," visible in an ophthalmologic exam, mark the fatty buildup.

By the first birthday, a child with Tay–Sachs disease withdraws. Vision fades. By two years, movement ceases, although the arms may suddenly strike out, a sign of profound brain damage. Swallowing becomes difficult, seizures worsen, and responses fade away. Children with Tay–Sachs disease rarely live past their fourth birthdays.

Mutations do not have a particular fondness for certain types of people. In fact, the first human genomes sequenced revealed that each of us probably carries at least twenty mutations that, if present in two copies, would cause disease. A mutation becomes affiliated with a group of people if the genetic change originates within the group, and then the people partner only among themselves, neither diluting the strength of the mutant allele in their gene pool as other people join, nor sending it out as people leave. Such inclusiveness may be by choice, or forced upon a people by the vagaries of history, as it was for the Jews.

The most common allele for Tay–Sachs disease among the Ashkenazim adds four bases to the gene, and this and two other mutations account for 98 percent of "Jewish" mutations. The mutations likely arose shortly after 70 AD, when the Jews were evicted from Rome and wandered, arriving in Russia and Poland by 1100 AD. Throughout history since then, as Jewish people were herded into ghettos and many perished, population bottlenecks (see Chapter 3) sampled, sequestered, and ultimately amplified the most common Tay–Sachs mutations. In addition, like carriers of cystic fibrosis and sickle cell disease, carriers of Tay–Sachs disease may have had an advantage that favored the mutant allele, such as resistance to tuberculosis or even a quick wit that enabled them to survive human-wrought horrors. A few other populations at elevated risk for Tay–Sachs disease have their own mutations—the Irish, Cajuns of Louisiana, the Old Order Amish in Pennsylvania, and French-Canadians.

The first clinician to describe Tay–Sachs disease, in early 1887, was Warren Tay, a dermatologist, ophthalmologist, and surgeon in London. Later that year, a New York City neurologist, Bernard Sachs, published a similar description. Sachs, an Ashkenazi who worked in the Jewish community, noted that cases were "almost exclusively observed in Hebrews." Tay reported, three years later, that his original case had since acquired two siblings, who were also sick. Four years after that, he noted the disease in families where the parents were related, suggesting autosomal recessive inheritance of an allele from a shared ancestor, such as a common great-grandparent. It wasn't until 1968 that the enzyme was discovered to be missing in children with Tay–Sachs disease, and in 1970 that carrier parents had a reduced level (O'Brien et al. 1971: 61). Could enzyme levels be used to identify carriers *before* they conceived affected children together? Michael M. Kaback, then a professor of pediatrics at Johns Hopkins University Hospital in Baltimore, devised such a test.

Carrier screening programs for Tay–Sachs disease soon began, targeting young married couples at synagogues and community centers, and students at colleges with large Jewish populations. Education was a major part of the effort. Students could meet with a genetic counselor at any time during the testing process, or years later, when carrier couples wanted to have children. Genetic counseling at that time was a brand new discipline.

The scope of the Tay–Sachs screening in the US in the early 1970s was staggering: 1.2 million young adults tested, 36,000 carriers identified, 1,056 carrier couples found, and 2,416 prenatal tests performed. As a result, disease incidence plummeted: 99 percent of identified carrier couples had children free of Tay–Sachs disease (Kaback et al. 1993: 2307). In the UK, the National Health Service began carrier screening in 1999. It focuses on schools and community centers in London and Manchester, where most of the UK's 270,000 Jewish citizens live.

In the early 1980s, an orthodox Jewish community in Boro Park, a section of Brooklyn, New York, took Tay–Sachs testing to a new level, with a program called *Dor Yeshorim* ("generation of the righteous" in Hebrew). Rabbi Joseph Ekstein founded it after losing four of his own children to Tay–Sachs disease. *The idea*: to test all teens in the community, with privacy protected, revealing carrier status when a couple begins seriously dating. *The goal*: to avoid carrier couple marriages, or to inform them so that they do not have children together. *Dor Yeshorim* is in accordance with the Jewish law not to end pregnancies. The "informed mate selection" strategy expanded both beyond the Brooklyn community and in the number of autosomal recessive conditions it covers.

By the 1990s, it became possible to switch from enzyme-testing for Tay–Sachs carriers to DNA testing. This new approach, however, actually introduced problems, because the types of mutation and their frequencies vary among populations. A test for the top three Jewish mutations, for example, would catch 98 percent of Ashkenazim, but only 20 percent of carriers in other groups, where other mutations are more common.

Another complication in administering DNA-based Tay–Sachs disease tests was marriage outside of the community, resulting in "admixture" of alleles. During the 1970s, admixture of Ashkenazim with many other groups increased from 13 to 43 percent. The "pure" Ashkenazi couples who had inspired Tay–Sachs testing were becoming rarer, and, as a result, DNA testing had decreased sensitivity in "mixed marriages." For these reasons, Tay–Sachs testing returned to tracking the phenotype—enzyme level—rather than the genotype —specific mutations (Gross et al. 2008: 54). However, for Ashkenazi couples, testing for specific mutations still makes sense. Several companies offer "Jewish genetic disease" panels. A person with four Ashkenazi grandparents has a one in four chance of being a carrier for at least one such disease. "Targeted genetic screening" of specific population groups is not discriminatory. Rather, it is based on the fact that some inherited diseases are more prevalent in certain populations, because of human behavior.

Another good example of successful targeted genetic screening is for beta thalassemia. This autosomal recessive anemia is caused by particular mutations in the beta globin gene, which is the same gene affected in sickle cell disease, discussed below. Beta thalassemia is treated with transfusions and drugs that counter the buildup of iron from the transfusions. This blood disorder is much more common in certain Mediterranean populations, such as in Greece, Sardinia, and Cyprus, than elsewhere in the world. In Sassari, northern Sardinia (part of Italy), for example, one in ten adults is a carrier (Longinotti et al. 2008: 238).

Several screening programs have educated communities about beta thalassemia, offered carrier screening and genetic counseling, and helped carrier couples deal with avoiding or treating the disease in their children. A program at high schools in Montreal, Canada, for example, offered testing for Tay–Sachs disease to Ashkenazi students and for beta thalassemia to students with Mediterranean ancestry. From 1972 through 1992, the program tested 14,844 Ashkenazim for Tay–Sachs disease, identifying 521 carriers, and 25,274 students of Mediterranean descent, identifying 693 carriers of beta thalassemia. Carrier couples identified when they were 16 years old returned for prenatal diagnosis years later. Incidence of both diseases fell by 90 percent over the two decades, and, today, most cases are outside the original high-risk groups (Mitchell et al. 1996: 793; McCabe 1996: 762). Critical to the success of the Montreal screen was that Canada has nationalized health care, and so people did not fear discrimination based on genotype.

SICKLE CELL DISEASE—A LESSON IN GENETIC DISCRIMINATION

Targeted population screening for Tay–Sachs disease nearly eliminated the illness. In contrast, targeted population screening for sickle cell disease caused confusion, fear, and stigmatization.

A specific, single-base mutation in the beta globin gene causes sickle cell disease, resulting in anemia and other symptoms. Recall from Chapter 2 that the hemoglobin molecule that transports oxygen in the blood is built of two alpha subunits and two beta subunits, encoded by two different genes (Figure 5.2). Under low-oxygen conditions, the abnormal hemoglobin proteins in a person with sickle cell disease align in a way that bends the red blood cells containing them into crescent shapes (Pauling et al. 1949: 543). These cells then lodge in tiny blood vessels, blocking circulation. This causes a stabbing or crushing pain in the body parts robbed of oxygen. An attack is called a "crisis."

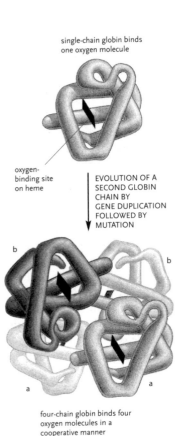

single-chain globin binds
one oxygen molecule

oxygen-
binding site
on heme

EVOLUTION OF A
SECOND GLOBIN
CHAIN BY
GENE DUPLICATION
FOLLOWED BY
MUTATION

b   b

a   a

four-chain globin binds four
oxygen molecules in a
cooperative manner

Figure 5.2 Globin. The hemoglobin molecule is built of four chains, two called alpha, and two called beta. Each globin chain holds a molecule of iron. Sickle cell disease and beta thalassemia affect the beta globin genes at different places

Source: *Molecular Biology of the Cell*, 5th edition, Figure 4–86, p. 256

Until the early 1980s, sickle cell disease was often fatal in childhood, from complications that included infection, stroke, pulmonary embolism, and kidney, liver, or heart failure. At that time, the name was changed from sickle cell *anemia* to *disease* to acknowledge additional symptoms. However, the disease is highly variable—a person might be only mildly ill, or need to be hospitalized often.

The sickle cell mutation arose independently multiple times in Africa, where the protection against malaria that it offered led to increased prevalence, as discussed in Chapter 3. The sickle cell mutation also arose in India or Saudi Arabia and is more prevalent among Mediterranean peoples. In the US, 1 in 400–600 African-Americans has sickle cell disease, and 1 in 10–12 is a carrier. In the UK, one in nine people in the high-risk population groups is a carrier—Africans, African-Caribbeans, Asians, and people from the eastern Mediterranean and Middle East. Among northern Europeans, sickle cell disease is very rare.

Reports from Africa of what was probably sickle cell disease date back to the 1600s. The first recorded case was Walter Clement Noel, a dental student in Chicago who had come from Grenada (Herrick 1910: 517). In 1904, Noel went to a hospital, complaining of great fatigue. A young intern, Ernest E. Irons, examined the student's urine and blood, did a physical exam, and took a medical history. The red blood cells in the smear were "peculiar, elongated and sickle-shaped," so odd that Irons sketched them and showed them to his superior, James B. Herrick. The two followed the dentist-in-training for the next few years as he suffered several sickle cell crises. Noel died at age 32 of pneumonia, and Herrick published his case without including his intern (Haller et al. 2001: 889). Another early case report described a seven-year-old girl with cough, night sweats, pain in her abdomen, legs, and joints, and fatigue, noting that her red blood cells sickled (Scriver and Waugh 1930: 375).

From 1970 to 1972, despite the lack of prenatal diagnosis, prevention strategies, or treatment for sickle cell disease, the US government began pursuing testing for the mutation. The goal was to identify carriers, who were unfortunately said to have "sickle cell trait." People with the disease would presumably already know it from symptoms. Testing was mandatory in 12 US states and indirectly targeted African-American children and young adults. For example, New York State law required sickle cell carrier testing of anyone applying for a marriage license not "of the Caucasian, Indian, or Oriental races" and of children who lived in cities—the center of the African-American community. Ironically, this was the time of the Civil Rights movement.

The term sickle cell "trait" was misleading (Fost and Kaback 1973: 742; Hampton 1974: 58). Carriers only rarely experience symptoms, which may happen when they are under low-oxygen conditions, such as at great altitudes. "Laws identified the diseases of 'sickle cell anemia' and 'sickle cell trait.' *Trait* is not the same as *disease*. The Air Force Academy wouldn't take recruits who had the trait. It became a real stigma," says Dr. Kaback, who led the Tay–Sachs disease screening program in the US and was heavily involved with sickle cell disease screening and testing too. At that time, when medical training included very little genetics, even some physicians were unaware of the distinction between "trait" and disease. The confusion persists. The Patient UK website, for example, lists prominent, brightly-colored adverts on the right side of the home page for "sickle cell trait," and clicking on it opens a page entitled "sickle cell disease information."

The sickle cell testing campaign in the US harmed some families. A major medical lab sold a test directly to consumers. A person would stick a finger, drip blood into a tube, shake it, and then hold the tube against a card that had lines on it. Obscured lines meant the person had some sickled hemoglobin molecules, but the test could not distinguish "trait" from "disease"—that is, the symptom-free heterozygote from the sick homozygote. "That little test spelled disaster. The Black Panthers and ministers, who meant well, bought boxes and boxes of the test and went door to door, testing little black kids. They would say, 'Mrs. J., your baby has sickle cell,'" recalls Dr. Kaback. Unconfirmed reports cited women committing suicide after learning that their children had "sickle cell," when the children really were carriers.

In 1972, the US Congress passed the National Sickle Cell Anemia Control Act, which tempered the discrimination. Prenatal testing became possible a few years later (Hollenberg et al. 1971: 698; Kan and Dozy 1978: 910). By the 1980s, mortality from sickle cell disease dropped, as affected newborns started a five-year course of anti-biotics and received immunizations to prevent deadly infections. Some patients benefited from bone marrow transplants or an anti-sickling drug, hydroxyurea. In 2004, the National Health Service in the UK began testing pregnant women, and then their partners if the women were carriers, and offering genetic counseling, pre-natal diagnosis, and pregnancy termination to carrier couples. By 2006, the NHS offered wider carrier testing, preconception genetic counseling, and newborn screening, focusing on areas in London and the West Midlands, home to population groups at highest risk.

The population screens for Tay–Sachs disease and sickle cell disease in the US differed profoundly, both in how the programs were conducted and in the nature of the two disorders. Tay–Sachs was the less variable and more severe condition—death invariably came in early childhood. At the time of the screen, prenatal diagnosis was possible, and the education campaigns stressed that carriers are symptomless. In contrast, sickle cell disease varies in severity, and, at the time of the screen, prenatal diagnosis was not possible. As Dr. Kaback says,

> The reason sickle cell shouldn't have been done is that there was no intervention (at that time). Find carriers, then the people were left hanging. The reason Tay–Sachs screening worked is that people had an option—not to have children, or prenatal diagnosis.

Public education was vastly different, as Dr Kaback explains:

> For 14 months we planned and developed the educational component for people to understand the idea of testing for Tay–Sachs carriers. The sickle cell programme was a disaster because there was no careful educational plan, and people misunderstood.

The historical and societal backdrop also differed profoundly: In the US at that time, Jewish people were better educated and had had more opportunities than African-Americans, and were less recognizable. As a result, sickle cell disease and trait became stigmatized; Tay–Sachs disease did not, and quietly vanished.

Dr. Kaback extends the Tay–Sachs and sickle cell experience to today's genetic tests:

> What we have learned from the history of genetic testing is that a test by itself is disastrous and not in the best interest of anyone. An absolute must is pre-test education, so people can understand what the test is for. Post-test counselling is also essential, because genetic diseases are relevant for family members.

Testing for Tay–Sachs disease and sickle cell disease helped pave the way for today's genetic tests. Table 5.2 lists current criteria for single-gene tests.

Table 5.2 Criteria for a population genetic test

- Excellent sensitivity (proportion of people diagnosed with a condition who actually have it; few false negatives)
- Excellent specificity (proportion of people found not to have the condition who actually do not have it; few false positives)
- Screening is less costly than not screening and paying for treatment
- Preventive measures and/or treatments available
- Pre-test and post-test genetic counseling offered
- Ethical review board approves the testing protocol
- Privacy maintained
- Informed consent required
- Outcomes clearly explained
- Program regularly evaluated

THE SECOND GENERATION OF GENETIC TESTS: DNA

The genetic screens of the 1970s detected specific proteins. By the late 1980s, when researchers had sequenced the genes behind several common single-gene disorders, DNA testing became possible (Lewis 1993: 14). However, because knowing the genotype (mutations) does not always predict the phenotype (traits or symptoms), results of a DNA test can cause undue anxiety and precipitate life-altering decisions. One reason for the disconnect between phenotype and genotype is penetrance, introduced in Chapter 3. Penetrance is the proportion of individuals inheriting a mutant genotype who actually have symptoms. Two genetic diseases—Huntington's disease (HD) and hereditary hemochromatosis (HH)—illustrate how penetrance affects the utility of a DNA-based test. For HD, penetrance is "complete": anyone who inherits a mutation and lives long enough will develop the disease. For HH, penetrance is "incomplete": only a small percentage of people who inherit a mutant genotype develop the condition.

HD: MOVEMENT AND MORE

Before the 1970s, HD was called "Huntington's chorea," which focused on its most obvious symptom, uncontrollable, dancelike movements.

"Chorea" became "disease" with the realization that subtle behavioral and cognitive (thinking) difficulties are part of the clinical picture. There isn't any curative treatment.

HD begins gradually and almost imperceptibly, usually in one's late thirties. The earliest signs may be obvious only to HD specialists —an odd positioning of facial muscles, or unusual eye movements. With hindsight, many people with HD realize that the first symptom was difficulty multitasking. Consider one woman's experience as a police dispatcher: "I couldn't answer the phone and listen to the police radio at the same time. They moved me to a different department, to take minutes. I couldn't do that, either, it was too fast." She burst into tears at her helplessness in following a familiar cake recipe, covering her kitchen in flour, and froze in utter confusion at the supermarket when faced with a wall of different cereals to choose from. The woman realized that her difficulty in decision-making and her depression echoed her father's illness. She knew that she would probably experience personality changes, such as becoming angry and irritable, and, ultimately, dementia.

Motor symptoms of HD usually begin shortly after cognitive clues. Fidgeting becomes more frequent and progresses to larger, repetitive movements. Eventually, hands wave and feet tap, in endless, apparently purposeless patterns. Loss of the perception of the body's location in space triggers tripping, falling, and colliding, as if intoxicated. Symptoms worsen. At first, the brain suddenly blocks grasping movements, and reached-for objects crash to the floor; later on, a grasp may morph into a powerful kick or lurch. Woody Guthrie, the famed folksinger, who died of HD in 1967, was overcome with twitches and staggers, constantly whirling about, injuring himself with his flailing limbs.

Like the woman who could no longer multitask, Woody Guthrie saw in himself the signs he knew all too well from his mother. That is the essence of autosomal dominant inheritance: it affects both genders, and every sick person has a sick parent. About 30,000 people in the US have HD, and about 8,000 in the UK. Many more people are "at risk," facing the one in two chance of having inherited the mutation from an affected parent. HD varies in age of onset, the sequence of symptoms, and their severity, even within a family. What doesn't vary is the inevitability of the disease. Therefore, a DNA test is highly accurate and predictive, as symptoms usually do not begin until adulthood. A healthy person who knows he or she has inherited the HD mutation is said to be "pre-manifest."

The HD mutation is unusual. It involves extra copies of a three-base DNA sequence in a gene that encodes a protein called huntingtin. The more extra copies, the earlier the onset, and the more severe the symptoms. The abnormally elongated huntingtin protein affects spiny-shaped neurons in a part of the brain that controls movement, preventing them from sending signals. This initiates a complex cascade of malfunction that kills the cells, lifting controls over movement.

An interesting history lies behind the development of the DNA test for HD. The earliest references to the disease concern women burned at the stake in Salem, Massachusetts, in 1692, after their peculiar movements led people to believe that they were witches. The first clinical description came in 1872, from "horse and buggy doctor" George Sumner Huntington (Huntington 1872: 317). As a boy in eastern Long Island, New York, George accompanied his father and grandfather, also physicians, as they visited patients, including two rail-thin women who contorted constantly, and men who walked as if drunk. Huntington presented his observations on this "hereditary chorea" at a medical meeting, noting the late onset and mental disturbance. Stating that the "thread is broken" after a generation does not have the disease, he described autosomal dominant inheritance three decades before Mendel's laws were rediscovered. He and others reported on more than 1,000 cases, over 12 generations of the family, tracing the mutation to two brothers from Suffolk, England.

Other large families helped researchers in the quest to discover the HD gene. In 1979, Columbia University psychologist Nancy Wexler, whose mother had HD, began leading yearly visits to a village where the homes are built on stilts, along the shores of Lake Maracaibo, Venezuela. Here, hundreds of people have HD. They are all descendants of a Portuguese sailor who brought the mutation to the settlement ten generations ago. The people traded Dr. Wexler blood samples for jeans and sweets. Searching the DNA in the blood samples for sequences that only the affected individuals shared led the investigators to a genetic marker—a sequence linked on the same chromosome as the HD mutation (Gusella et al. 1983: 234). All the sick Venezuelans had it. Because a single gene causes HD, the marker would help other families, too.

By 1986, the genetic marker test became available to other families with HD, with limitations. Several relatives had to participate to track the inheritance pattern, because, in some families, the marker would not lie on the same copy of chromosome 4 as the HD mutation itself. This is why accuracy of the marker test was 96 percent, and not 100 percent. Finding the gene itself, a decade later, made the marker test obsolete, increased accuracy, and made a test possible for anyone— no relatives required (Huntington's Disease Collaborative Research Group 1993: 971).

Paramount in planning and delivering the HD test was considera-tion of the complex emotional and bioethical issues associated with predicting disease in a healthy, young person. The test could tell whether or not an individual could pass the disease to their children. In the past, at-risk individuals had children without knowing this, because symptom onset is usually in one's late thirties. Intensive, months-long genetic counseling determined whether a person was psychologically prepared to handle the consequences of testing. The HD test is always given in a medical setting, such as the UK's National Health Service's genetics centres (Harper et al. 2000: 567). Uptake of the test has been less than researchers expected—in many nations, only 15–25 percent of at-risk individuals take it. Fortunately, most people who have dealt with the information that the test provides have done well. People take the test to end the uncertainty and to help with making decisions about having children, choosing education and careers, financial planning, and other actions.

HEREDITARY HEMOCHROMATOSIS: IRON OVERLOAD

In the best of genetic tests, genotype predicts phenotype, as it does for HD. A genetic test is less useful if penetrance is low, leading to false positives—mutations are there, but the person may not get sick. This is the case for hereditary hemochromatosis (HH).

HH causes "iron overload." The bloodstream absorbs too much iron from food and, over many years, deposits the excess in vital organs, such as the liver, pancreas, and heart. Skin color changes, as noted in the initial description of the disease in a 28-year-old: "From the time this man came into the hospital, I was struck by the almost bronzed appearance of his countenance, and the blackish color of his penis" (Trousseau 1865: 663). If symptoms are untreated, the disease is deadly. Fortunately, treatment is easy: remove blood periodically to keep the iron level down. Menstruation takes care of this naturally for women, who typically do not show symptoms until after menopause.

Two common mutations in a gene called *HFE* cause most cases of HH, which is autosomal recessive. Before the gene was sequenced in 1996 (Feder et al. 1996: 399), the diagnostic test measured how strongly a protein called transferrin binds iron. Adding DNA testing would presumably add precision, but, owing to incomplete penetrance, the opposite occurred. As was the case for Tay–Sachs disease, detecting the phenotype (extra-strong transferrin) was more meaningful than determining the genotype. One study found that the percentage of people who inherit mutant genotypes and who actually develop HH symptoms is less than 1 percent (Beutler et al. 2002: 210)! Penetrance varies by population, but for HH it is always low—and that greatly limits the utility of genetic testing.

HH *is* worth testing for, however, because it is common, and complications are preventable. In most European populations, 10 percent of people are carriers; in Ireland it is 20 percent. Tests for HH offered direct-to-consumer on the Internet are confusing, because they provide little or no information about penetrance. For example, the American Hemochromatosis Society webpage urges DNA testing: "All it takes is a . . . DNA test kit . . . and you can confirm, or rule out, Hereditary Hemochromatosis in a matter of days." That is not so, because most people with the mutation do not actually have the disease. The test can cause unnecessary anxiety.

The testing programs for all of these single-gene disorders have taught us that it is important to understand the biological basis of a disease to best use the information. This is not often the case for mail-in DNA tests offered on the Internet, and consumers unfamiliar with genetics may be misled.

THE THIRD GENERATION OF GENETIC TESTS: DO-IT-YOURSELF

Direct-to-consumer genetic tests include both proven, single-gene tests, as well as "associations" of genetic markers with medical conditions. The most commonly used type of genetic marker is a SNP, which is a site in the genome where the base differs in more than 1 percent of a population (see Chapter 4). In a genome-wide association study (GWAS), researchers compare thousands to millions of SNPs in thousands of people with a condition, and an equal number of people without it, but matched in other ways, such as gender and age. A computer algorithm detects patterns exclusive to the sick people, and the sites of the shared SNPs suggest where to look among the chromosomes for genes that contribute to the risk of developing the condition. However, genes detected in these studies often contribute only a very small degree to the overall disease risk. Perspective is important for interpretation too. For example, cigarette smoking may raise lung cancer risk much more than inheriting a particular SNP pattern. These studies are very helpful in research, but the validity of applying them to individuals isn't yet known. Still, Internet-based companies offer them.

NUTRIGENETICS TESTS

Some direct-to-consumer genetic testing companies specialize. This is the case for "nutrigenetics" companies that sell nutritional supplements supposedly based on a customer's genotype. An investigation by the US Government Accountability Office (GAO) exposed misleading interpretation of test results from four nutrigenetics companies (www.gao.gov/cgi-bin/getrpt?GAO-06–977T). A consumer sends in a DNA sample collected with a swab from the inside of the cheek, and the company identifies specific alleles or SNPs, then matches the results to "dietary and lifestyle" suggestions. Test interpretation is based on population-level data.

The GAO researchers took DNA from a nine-month-old girl, a 48-year-old man, a dog, and a cat, and submitted the samples with invented diet/lifestyle profiles representing 14 people: 12 female and 2 male. For example, the baby girl DNA was submitted as coming from a 45-year-old man, who is 1.8 meters (6 feet) tall and weighs 95.3 kilograms (210 pounds), smokes and doesn't exercise, drinks a lot of coffee, and eats a fatty diet. The baby girl also masqueraded as a 72-year-old woman, who stands 1.4 meters (4 feet, 9 inches) tall and weighs 45.4 kilograms (100 pounds), regularly exercises, never smoked, doesn't drink coffee, and eats fried foods. The companies presented what one of them terms a "personalized program based on your unique genetic make-up."

Results revealed that the companies provide anything but personalized information. Certain, very different "fictitious consumers" learned that they were at increased risk of developing osteoporosis, high blood pressure, type 2 diabetes, and heart disease—each of which could result from any of many combinations of alleles and environmental factors. However, some "customers" who submitted the same DNA nevertheless learned that they had different genetic conditions. All customers were offered nutritional supplements for a price inflated 100-fold over cost at a grocery. Advice dispensed was both generic and obvious: eat vegetables, don't smoke, and exercise. The GAO report concludes: "Although these recommendations may be beneficial to consumers in that they constitute common sense health and dietary guidance, DNA analysis is not needed to generate this advice."

PERSONALIZED MEDICINE: GENETIC RISK TESTS

In health-based, direct-to-consumer genetic testing, a person sends in a cheekbrush or saliva sample and, a few weeks later, receives absolute risk assessments for chosen conditions based on identifying genetic markers—DNA sequences associated with a particular phenotype. Such companies disguise their offerings by calling them "informational," with disclaimers such as "this is not a genetic test for disease or predisposition to disease, nor does it determine a medical condition." Still, consumers may interpret DNA test results as having the same validity as a doctor's diagnosis. With this in mind, J. Craig Venter, of human genome sequencing fame, and colleagues investigated the accuracy of health-based tests in an analysis similar to the nutrigenetics probe (Ng et al. 2009: 724). They submitted DNA samples from five real adults to two companies, both based in California—23andMe and Navigenics—and requested risk figures for 13 diseases. Table 5.3 lists the health conditions.

These diseases are multifactorial: caused by any of several genes and environmental factors. This means that any one genetic marker may contribute only slightly to a particular condition—that is,

Table 5.3 Disorders tested for in investigation of 23andMe and Navigenics

Breast cancer
Celiac disease
Colon cancer
Crohn's disease
Heart attack
Macular degeneration
Multiple sclerosis
Prostate cancer
Psoriasis
Restless legs syndrome
Rheumatoid arthritis
Systemic lupus erythematosus
Type 2 diabetes

inheriting it is not a diagnosis, but a small, theoretical elevation in risk. More often than not, the extent to which a marker contributes to a specific disease risk isn't even known. A company's test for a particular condition might not include all responsible genes. One woman's test results, for example, claimed that, according to her DNA, she was at very low risk of developing diabetes—a disease that she and other family members had had for many years. The company never considered family history information in their analysis.

The source of uncertainty in the tests that Venter and his associates tested was not in the DNA sequencing, but in the risk analysis. Relative risk is comparison with others; absolute risk is for a particular individual. Absolute risk is the relative risk multiplied by the average population disease risk. For example, if a woman's genetic markers indicate a 1.5 relative risk for emphysema, this means that her risk is 50 percent higher than average. The average population disease risk for emphysema is 10 percent. Therefore, the woman's absolute risk is 15 percent (1.5 × 10 percent). For a condition with a very low average population disease risk—for example, one in 10,000—the absolute risk is vanishingly small and practically meaningless.

Both relative and average population disease risk can be inaccurate. If a marker–disease association has not been tested in a very large number of people, the predictive value may be low. The average population disease risk is also prone to error, because different companies define populations differently. A customer who is Greek–Chinese might have little to learn taking tests from a company whose reference population is northern European. Navigenics defines populations as male or female; 23andMe uses age cohorts. These differences explain why companies give different absolute risks for the same DNA samples. Another source of disagreement is when companies use different markers for the same diseases.

These inconsistencies in risk assessment explain why, for 7 of the 13 diseases in the Venter study, the companies' test results significantly differed. Those diseases that agreed had a very strong risk allele. For celiac disease, for example, the risk allele that both companies used is present in 90 percent of cases. Therefore, finding it is reliable information.

Venter and his colleagues suggest ways that direct-to-consumer tests can improve:

- Report the genetic contribution for each marker.
- Use only high-risk markers.
- Evaluate markers with population data appropriate for customer ancestry.
- Investigate how customers interpret and use genetic risk information.
- Conduct prospective studies to see if risk predictions are accurate.

People interested in learning about their inherited disease risks need not spend anything—they can participate in research projects. For example, the Coriell Personalized Medicine Collaborative, based at the Coriell Institute in New Jersey, provides risk estimates for common, multifactorial diseases and tracks how people use the information.

POPULATION BIOBANKS

Genetic testing is growing, both in the numbers of different tests and the numbers of test-takers. It is being conducted at the population level, too. Several countries maintain population biobanks, which store genetic as well as lifestyle, medical, epidemiological, environmental, and genealogical information. The idea is to develop a database large enough to reveal the contributing factors to specific health conditions.

The first population biobank, established in 1998 by a government-backed company called deCODE Genetics, used the excellent genealogical records of Iceland. Researchers at deCODE have discovered many important genes that contribute to common, multifactorial disorders. In the US, the NIH is planning a genetic–epidemiologic biobank that will collect personal information, including genotypes, from half a million people, including 120,000 children. The study will mine the data for associations among genotypes, environmental factors, and health conditions. Other variations on the biobank theme target specific populations. The UK Biobank is looking at genes, lifestyle habits, and health in half a million people between the ages of 45 and 69, when symptoms of many multifactorial conditions appear. The Estonian Genome Foundation uses data from patient registries for cancer, osteoporosis, diabetes mellitus, and Parkinson's disease.

DNA sequences hold information other than health predictions. Certain sequences of A, T, G, and C represent our identity and uniqueness. Therefore, privacy protection has been necessary, as genetic testing has become more widespread. For example, the Genetic Information Nondiscrimination Act, enacted in the US in 2008, protects against discrimination in health care and in the workplace based on genotype. Other uses of DNA testing are to provide ancestry information, discussed in Chapter 1, and forensics.

DNA PROFILING

On September 4, 1993, seven-year-old Ashley Estell vanished in a Texas park when she wandered away from her parents, who were watching her brother play soccer. The next day, her body was found nearby, strangled. Evidence pointed towards a man named Michael Blair. Several people had seen him at the football game, he had once committed a sexual crime, and his car was found near the body. Although he claimed the car was there because he was helping to find her, he was arrested. Hairs found in his car were used as evidence and compared with hairs on the victim and found at the park. At the trial, testimony focused on the eyewitnesses (who had never actually seen the accused with the victim) and analysis that found the hairs "microscopically similar." Blair was sentenced to death and served 14 years, before DNA testing proved that the evidence was not from him. In 2002, DNA tests were run on the hairs and on material from under the victim's fingernails—and none of it matched Blair. The Innocence Project, formed in 1992 to help the wrongly convicted use DNA testing to prove their innocence, helped Blair.

A DNA profile is a test that compares DNA from different sources, such as at a crime scene and from a suspect. It represents several places in the genome that vary among individuals. Sir Alec Jeffreys developed the technique in the 1980s at Leicester University in the UK. He distinguished individuals by the numbers of copies of certain short repeated DNA sequences. Multiplying the frequencies of the different repeat numbers in the suspect's population group provides the probability that two sources of DNA are alike by chance. If this figure is very small, then a conviction is likely. For example, in the US, the Federal Bureau of Investigation's Combined DNA Index System (CODIS) considers 13 repeated sequences, each present in varying numbers in different populations. The chance that any two non-related individuals in the same population group have the same number of repeats at all 13 sites is 1 in 250 trillion. So, if Michael Blair's DNA had matched the hairs found on the little girl's body at all 13 sites, it would be fairly certain that he had committed the crime. Just one difference in a DNA profile proves innocence.

Another application of DNA profiling is in identifying disaster victims. It was useful in identifying remains after the terrorist attacks in the US on September 11, 2001, after hurricane Katrina in 2005, and for the earthquake victims of 2010, but was less helpful in the tsunami of late 2004, because the evidence washed away. Like health-related genetic tests, DNA profile tests are also offered on the Internet, to test paternity and to provide evidence of infidelity.

Whether through government-run biobanks, direct-to-consumer websites, part of preventive health care, or in forensic investigations, it is likely that genetic testing lies in all of our futures—and certainly in those of people not yet born.

WHAT WOULD YOU DO?

The six-year-olds line up, set to race 100 meters so that the watching coaches can determine who is likely to be a better soccer player. The kids aren't nervous—they're excited and having fun—but some of the parents are.

"I'm not worried at all. Jared's guaranteed to be a great forward," said one mother, smugly, to another.

"Guaranteed? How can you know that?" asked the other parent, looking skeptical.

"A genetic test. Jared was tested when he was a baby, so we knew he'd be a superior sprinter. Pretty great, right?"

The other mother was quiet a moment. Then she spoke. "Not to sound dumb, but why couldn't you just wait until he got a little older, to see if he runs fast? And if he even *likes* running?"

"You just don't understand."

Actually, the mother who had a genetic test performed on her young son didn't understand. The test was for an allele of a gene called *actinin 3 (ACTN3)*. Two alleles of the gene are known, *R* and *X*. After a study reported that people with two *R* alleles (*RR*) are overrepresented among sprinters, and people with two *X* alleles (*XX*) are more common among endurance athletes, a company invented a test, guessing that parents anxious for their children to be the best would race to it. They were right.

Theoretically, the test is based on the fact that the *R* allele encodes a protein that functions with fast twitch muscle fibers, powering sprinting. When the gene is deleted (the *X* allele), the protein is not made, and the slow twitch muscles instead have an advantage, promoting endurance. Through evolutionary time, either genotype had an advantage, depending upon whether it was more adaptive for our ancestors to be able to sprint past a cheetah and leap into a tree, or outrun a carnivore over a long distance. Both alleles, therefore, have stood the test of time.

Practically, the *ACTN3* test smacks of genetic determinism, the idea that our genes define us. But our genes do not act in a vacuum. Other genes, as well as environmental influences, greatly impact athletic ability. One geneticist remarked of the test, "Why not simply line kids up and see who runs the fastest?"

Would you have the *ACTN3* test performed on your young child?

SUGGESTED READING

The literature on testing for genetic disease spans several decades and revisits the same set of issues based on use of genetic information. An excellent early work is *Proceed With Caution: Predicting Genetic Risks in the Recombinant DNA Era*, by Neil A. Holtzman (Johns Hopkins University Press, 1989). For blogs that compare the genetic tests for the well-studied, single-gene conditions discussed in this chapter with direct-to-consumer genetic testing, see, by Ricki Lewis:

- http://blog.bioethics.net/2008/04/the-gap-is-widening-on-genetic-testing-too/
- www.scienceprogress.org/2008/05/a-brief-history-of-genetic-testing/

Alice Wexler, whose mother died of Huntington's disease, has written two excellent and highly readable books about the history of the condition: *The Woman Who Walked into the Sea: Huntington's and the Making of a Genetic Disease* (Ann Arbor, Michigan: Sheridan Books, 2010) and *Mapping Fate: A Memoir of Family, Risk, and Genetic Research* (Random House, 1995).

The standard textbook on DNA profiling is *Forensic DNA Typing*, 2nd edition, by John M. Butler (Elsevier Academic Press, 2005). For one of the earliest articles on DNA profiling, then called DNA fingerprinting, see the cover story for the June 1988 issue of *Discover Magazine*, "DNA fingerprints: Witness for the prosecution," by Ricki Lewis.

6

GENETIC AND GENOMIC TECHNOLOGIES

Over the past half century, the field of human genetics has evolved from a single-gene to a multifactorial approach, spawning valuable technologies along the way. In the 1950s and 1960s, researchers matched human genes to specific health conditions, with the help of families in which an abnormal chromosome tracked with the disease. In the 1970s, recombinant DNA technology gave microbes human genes, which produced protein-based drugs. In the 1980s, sequencing the human genome was an idea; in the 1990s, it became reality. Since then, we have been seeking meaning in the data.

Mining the mountains of DNA sequence data required a new field, **bioinformatics**, which weds computer science with biochemistry. Bioinformatics has mushroomed with the new millennium, as the human genome has turned out to be much different than anyone expected. Francis Crick once envisioned that knowing the genome of an organism would be to know all there was to know about it. He was wrong. Instead of being like a book that tells a simple, linear story, the human genome is, instead, a source of information that is both shaded and nuanced, dynamic and difficult to predict, filled with subtext to the seemingly simple sequences of A, C, T, and G. The human genome in a person may be ephemeral, but, overall, is a timeless creation of nature, at the same time a repository of our past as well as the raw material for our future. In the present, however, we can use gene and genome information to improve our health and to learn more about the other species with whom we share the planet. This final chapter examines three DNA-based technologies, progressing from the single-gene level (*gene therapy*), to suites of interacting genes (*gene expression profiling*), to viewing the planet as a place packed with genomes representing many types of organism (*comparative genomics*).

GENE THERAPY

Imagine a recipe that has an error. It calls for too much or too little of an ingredient, leaves one out, or perhaps includes something not quite right. Correcting the instructions can perhaps salvage the dish. This is the idea behind gene therapy: replacing faulty genetic instructions with the normal form of the gene.

HOW GENE THERAPY WORKS

Since 1990, more than a thousand gene therapies have been through clinical trials. However, the approach has yet to graduate from the experimental stage. This is because the hurdles are high. The replacement gene must correct a problem in enough affected cells, for long enough to help, without harming other cells. That's a tall order.

The first step in developing a gene therapy is to select a disease that meets certain criteria. The mechanism should be understood, the causative mutation known, the affected body part accessible, and available treatments inadequate or non-existent. The second step is to design a way to deliver the healing DNA. Bombard or inject a cell with DNA? Ferry it in inside tiny fatty bubbles, called liposomes? Or stitch the human gene into the genetic material of a virus and then infect the target cells with the engineered viruses? The viral route is the best.

Today, most gene therapies use the restriction enzymes of recombinant DNA technology to excise the healthy version of a human gene and then insert it into a virus that serves as a "vector." Although we commonly view viruses as causing disease, in gene therapy, their simplicity makes them ideal gene-carriers—they are little more than genetic material wound into a protein shell. Take out the parts of a virus that cause disease and add a human gene of interest, send it into the right type of cell, and that is a gene therapy. The strategy takes advantage of viruses' natural tendency to inject their genetic material into host cells, where they commandeer the protein synthesis machinery to make more of themselves. We are asking them to do what they normally do, while carrying a swapped-in, stowaway human gene. Researchers select a particular type of virus based on the virus's natural target. Some viruses infect only certain cell types; others infect only cells that are dividing. Were the genetic code not universal—all species and viruses manufacture the same amino acids using the information in the same RNA triplets—gene therapy would not be possible.

For gene therapy, the path from theoretical idea to practical treatment has been like a roller coaster, with tantalizing highs followed by plunging lows. After initial successes, too many clinical trials ended in short-lived improvement, or none at all. Then, in 1999, a death due to gene therapy stalled the field until a decade later, when a handful of promising results revived interest. Below are the stories of some of the children who have helped to put gene therapy on the map. Table 6.1 summarizes their conditions.

Table 6.1 Gene therapies discussed in this chapter

Disease	Mode of inheritance	Symptoms
ADA deficiency	Autosomal recessive	Immune deficiency
Ornithine transcarbamylase deficiency	X-linked recessive	Coma
Leber congenital amaurosis	Autosomal recessive	Nightblindness, eventual total blindness
Adrenoleukodystrophy	X-linked recessive	Social withdrawal, weakness, skin darkening, seizures, deafness, blindness, poor coordination, dementia

INITIAL GENE THERAPY SUCCESS

The first gene therapy to be tested treated a rare form of immune deficiency called *adenosine deaminase (ADA) deficiency*. Lack of an enzyme called ADA destroys helper T cells, which are needed to activate the B cells that produce antibodies. (Chapter 4 introduced the parts of the immune system and how they interact.) Crippling helper T cells shuts down the immune response. Before the 1990s, children with ADA deficiency typically spent their short lives in and out of hospitals, receiving blood transfusions and bone marrow transplants to temporarily bolster their failing immunity.

The prelude to gene therapy for ADA deficiency was enzyme replacement therapy, in which children received weekly infusions of the enzyme. This approach worked, but researchers knew there was a better way. Replacing the *gene*, rather than the *protein*, would provide a longer-lasting, and perhaps permanent, source of ADA. Best of all, the missing enzyme would come from the child's own body.

The first gene therapy for ADA deficiency was delivered on a September afternoon in 1990, at the NIH in the US. Four-year-old Ashanthi DeSilva received an infusion of her own T cells, removed earlier and given normal ADA genes. It worked, but only transiently, for T cells have a short lifespan. Ashanthi and another little girl who was treated, Cynthia Cutshall, improved, but they still had to take enzyme supplements to maintain their immunity. Researchers wondered if altering a different type of cell in bone marrow, a cell that gives rise to blood cells, would provide a longer-lasting effect. To test this idea, in 1993, three newborns, diagnosed before birth with ADA deficiency, received their own umbilical cord stem cells, with the healthy genes added. Because these cells continually gave rise to T cell precursors, immunity lasted longer, and a greater percentage of the babies' blood cells had corrected genes than had been the case for Ashanthi and Cynthia.

Gene therapy efforts continued, with more disappointments than Eureka! moments. For some disorders, effects were too limited to have a clinical impact. Delivering normal CFTR proteins to the air passages of a person with cystic fibrosis would improve breathing for a few weeks—maybe. A gene therapy for Duchenne muscular dystrophy worked, but only at the exact points where the genes were introduced—big toes. At the same time, the human genome project was gearing up, and the news media hyped the expectation that, somehow, learning our entire DNA sequence would reveal ways to cure diseases. When the genome project was just two years from completion, in 1999, progress in gene therapy stopped, with the death of a teenager.

GENE THERAPY FAILURE

Jesse Gelsinger was born in 1981. Shortly before his third birthday, following a mild respiratory infection, he suddenly became belligerent and aggressive. As his mother had a history of schizophrenia, the worried parents took him to the doctor. Suspecting simple fatigue from not eating well, the doctor placed Jesse on a high-protein diet— a decision that nearly killed the boy. Within a few days of being forced to drink milk and eat peanut butter, Jesse fell into a deep sleep. When his parents couldn't rouse him, they put him in the car and drove over the bridge from New Jersey to the Children's Hospital of Philadelphia. Jesse was in a coma.

After a week of tests, the Gelsingers learned that Jesse had an inborn error of metabolism called *ornithine transcarbamylase (OTC) deficiency*, which is recessive and inherited on the X chromosome. Jesse actually had a very mild case, because he was a genetic mosaic, with the mutation in only some of his cells. The mutation must have occurred when he was a two- or perhaps four-celled embryo, in just one of the cells, so that half, or a quarter, of the embryo that would develop into Jesse had OTC deficiency. It was enough of an enzyme deficiency to make him very sick sometimes, but it would likely not have killed him.

The OTC deficiency impaired Jesse's ability to digest protein, which is why he became so ill following the doctor's advice to pack in the peanut butter. The enzyme normally liberates the nitrogen from proteins, excreting it as urea in urine. In OTC deficiency, the freed nitrogen bonds with hydrogen, forming ammonia (NH_3). This chemical is as harsh in the body as it is on a dirty floor. It builds up in the bloodstream and travels to the brain, where it does great damage. Most babies born with the mutation in every cell become comatose the first day and die by the second. For those who live with the disease, such as Jesse, following a very-low-protein diet and taking medications that bind ammonia can prevent symptoms.

However, this regimen takes a lot of self-control. A parent can keep a young child on the program, but an adolescent might let the diet or drug schedule lapse. That is what happened to Jesse.

In 1998, when Jesse was a senior in high school, a busy school and work schedule kept him from taking his pills on time, and his health began to fail. That December, he became acutely ill and fell into a coma, from which he nearly didn't awaken. His father, Paul, had heard about a gene therapy trial at the University of Pennsylvania in Philadelphia and, alarmed at the illness that was barely under control, mentioned it to the usually rebellious Jesse. Much to the father's surprise, Jesse was interested! They were both excited at the possibility of participating, but, when Paul called to get further information, he was disappointed to learn that Jesse would have to wait until his eighteenth birthday. So that is what Jesse did.

On Monday, September 13, 1999, Jesse became the eighteenth person in the clinical trial to receive billions of a type of virus called an adenovirus (AV) in his liver. Each virus had, as part of its DNA, a copy of a normal human OTC gene. But, instead of entering the type of liver cell that would correct the genetic defect, the viruses entered immune system cells in the liver, triggering a massive, body-wide immune response (Figure 6.1).

The first night, Jesse had a fever, but this was not an unusual reaction to receiving a huge load of viruses. However, by the next day, the ammonia level in his blood had shot up, as red blood cells burst, releasing hemoglobin protein and the nitrogen, which his body then turned into the toxic ammonia. He failed fast. By the time drugs lowered the ammonia level, it was too late—Jesse's blood had stopped clotting, and his organs were shutting down. By Friday, he was declared brain dead, and Paul turned off life support. The autopsy revealed that Jesse had a pre-existing, undetected parvovirus infection, to which he had built up immunity. When huge numbers of AVs had suddenly flooded his system, his immune response was primed and ready. Instead of protecting him, it swiftly killed him (Gelsinger 2001).

Jesse's death pointed out the failure to assess adequately the immune status of a participant in a clinical trial for a gene therapy. This prompted revamping of clinical trial protocols in the US, and only a few experiments were permitted to continue. However, further trouble lay ahead. In 2000, researchers in France published the first

Figure 6.1 Jesse Gelsinger. Jesse Gelsinger had just turned 18 years old when he died a few days after receiving an experimental gene therapy. In this photo, he is standing on the stairs at the Philadelphia Museum of Art, where Rocky Balboa stood in triumph in the film "Rocky." Jesse's bravery ultimately led to safer gene therapy

of several reports of success in treating nine boys for an inherited immune deficiency, also with virally delivered healing genes (Cavazzana-Calvo et al. 2000: 669; Aiuti et al. 2002: 2410). These experiments used a retrovirus, which has RNA as its genetic material that is copied into DNA during infection. Although the boys recovered from their inherited disease, three of them developed leukemia as a direct result of the gene therapy—the viruses had integrated into oncogenes (discussed in Chapter 4). One boy died from the leukemia.

The deaths of Jesse Gelsinger, the boys with leukemia, and then, in 2007, of a young woman being treated with a gene therapy injected into one knee to fight arthritis, sidelined many promising gene therapy trials. The picture changed in late 2009.

GENE THERAPY SUCCESS AGAIN

On September 28, 2008, an eight-year-old boy lay immobilized on a table at the Children's Hospital of Philadelphia, as a surgeon carefully placed a few billion viruses into his left eye. Each virus carried a working copy of a human gene that was mutant in both copies in the boy, making him legally blind, living in a world of shadows. Corey Haas was genetically destined to be completely blind by early adulthood.

Four days later, Corey and his parents were visiting a zoo near the hospital when he suddenly put a hand over his treated eye, complaining that the sunlight was bothering his eyes. This had never happened before. A month later, back home in a small town in the Adirondack Mountains of upstate New York, Corey was raking leaves with his father, Ethan. The boy looked up, and his father was stunned. "His eye was so blue! I'd never seen the blue in his eyes before," Ethan Haas said. Corey's pupils had constricted in the bright sunlight, revealing the blue irises that matched those of his father.

Corey had inherited a different recessive mutation for *Leber congenital amaurosis* (LCA) from each parent, making him a "compound heterozygote." There was no family history of visual problems, but the family was small, and the problem recessive; the responsible allele passed silently along the main branches of his family tree. The babysitter was the first to suspect poor vision. Corey did not make eye contact and would stare at light bulbs, the intense illumination apparently not bothering him. By the time Corey was walking, a visual deficit became obvious—he constantly crashed into things. And he still stared at the lights.

Pediatricians were puzzled. Then, an educator who watched the boy was reminded of another child with the same symptoms being helped by a specialist in Boston. Corey went there too. Over the next few years, the doctors narrowed down the diagnosis to LCA, and a genetic test then diagnosed the subtype. Corey was extremely lucky—a gene therapy clinical trial for his exact subtype of this rare disease was about to start. Much of the research had been done on a breed of sheepdog that has the same condition, and gene therapy had already cured a dog named Lancelot, who had visited the US Congress to show off his progress (and lobby for funding). Corey joined the clinical trial and, today, he and 11 others, their vision partially restored, await treatment on their second eyes (Amado et al. 2010: 1).

Corey's disease, LCA, was a candidate for gene therapy because researchers knew exactly what goes wrong. At the back of the eye, a layer of tile-like cells, called the retinal pigment epithelium (RPE), supports and removes wastes from the rods, which are the "photoreceptor" cells that supply black-and-white vision. A protein in the RPE normally turns vitamin A into a form that can nourish the rod cells. Without this protein, Corey's rods were slowly dying, and, with them, his vision in dim light was ebbing. The healing genes sent in with the viruses had entered his RPE cells and were producing the protein, enabling him to see. The gene therapy for the eye disease, together with many other ongoing clinical trials for gene therapy, uses adeno-associated virus (AAV), which is safer than the AV used on Jesse Gelsinger.

Also in 2009, researchers in France reported effective gene therapy for adrenoleukodystrophy (ALD), an inborn error of metabolism that was featured in the 1992 film Lorenzo's Oil (Cartier et al. 2009: 818). ALD is inherited on the X chromosome. Boys show symptoms between the ages of five and eight and rarely survive their tenth year. Like many genetic disorders, ALD symptoms are many and seemingly unconnected. The first signs are, typically, social withdrawal and difficulty concentrating. The boy may become weak, and his skin may darken. He eventually suffers seizures, deafness, blindness, loss of coordination, and dementia.

The problem in ALD is a stripping of the fatty insulation, called myelin, from nerve cells in the brain and spinal cord. (In Tay–Sachs disease, described in Chapter 5, neurons become buried in *too much* myelin.) The defect in ALD is in the cells, called neuroglia, that wrap around the neurons, providing the lipid coat that speeds nerve transmission. The enzyme that is missing normally escorts an enzyme into the peroxisomes, the tiny sacs in cells where used lipids are dismantled to recycle their parts.

The film *Lorenzo's Oil* was based on the true story of Michaela and Augusto Odone, who taught themselves the biochemistry behind their son Lorenzo's ALD and devised a mix of olive and canola oil ingredients that theoretically should stall progression. It did, but unfortunately Lorenzo's brain had already sustained too much damage by the time he started taking the oil. Still, he lived until age 30, rather than dying in boyhood as his doctors had predicted. Even with the oil, Lorenzo was bedbound and unable to speak. Today, some boys with ALD benefit from a stem cell or bone marrow transplant from a matched donor. However, these transplants are dangerous and have side effects, and many boys cannot find donors. As researchers understood the disease so well, it was a good candidate for a gene therapy.

The successful gene therapy for ALD removed the boys' own bone marrow stem cells, patched them with wild type alleles, and reinfused them. The virus used to deliver the gene may seem surprising—it is HIV, with the genes that cause AIDS removed. The researchers chose HIV because it readily infects non-dividing cells such as neuroglia, and it does not insert into oncogenes, as the viruses used to treat the boys in France in 2000 did. The HIV approach has halted disease progression in three boys with ALD, again in France. Brain scans show that myelin is being produced again.

Will the gene therapy successes of 2009 place the field back on track? It has backslid before, and so the future remains uncertain. In the meantime, many more people may benefit from another genetic technology, gene expression profiling.

GENE EXPRESSION PROFILING

A **gene expression profile**, or signature, is the collection of proteins that a particular cell manufactures under specific conditions.

It provides a more dynamic view of a cell, complementing such static descriptors of classical histology as shape, size, and staining properties. Gene expression profiling is becoming a very useful tool in studying cancer, in basic research as well as in the clinic. The technique is valuable at all stages, from initial diagnosis, to selecting treatment, to tracking response to treatment, and in predicting the likelihood of a cancer's spread or recurrence.

A NEW LEUKEMIA

Leukemia is a cancer of the white blood cells that causes fatigue, easy bruising, and fever. These symptoms result from an overabundance of white blood cells, which crowd out other cells in the bone marrow, the tissue that gives rise to blood components. The resulting deficit of red blood cells produces the fatigue, and deficiency of platelets causes the bruising.

When chemotherapy became available for childhood leukemia in the 1970s, it was a resounding success—many children survived their cancers. At the time, oncologists recognized two major acute (sudden onset) types of the disease based on the affected cell type—acute lymphoblastic leukemia (ALL) and acute myeloid leukemia (AML). Chemotherapy helped about 90 percent of patients with either of these two types of acute leukemia. The other 10 percent would not respond at all, or would do so for a short time and then relapse, most dying before their first birthdays. In 2001, one of the first applications of gene expression profiling finally explained the deaths. The unresponsive 10 percent were not just unlucky; they had a different, third form of the disease. Children with any of the three types of leukemia have the same symptoms, and in fact their cancerous cells look the same under a microscope. But, on the molecular level, they differ.

The first inkling of a third form of leukemia came in 1979, when researchers noted that a small percentage of leukemia patients had cells with chromosomes broken in a specific way: part of chromosome 11 had exchanged places with part of chromosome 4. Then, in 1984, the invention of fluorescence activated cell sorting, which uses a laser to distinguish cell types by their surface molecules, enabled researchers to see that a few leukemia patients had cancer cells that bore surface characteristics of *both* ALL and AML (Slany 2009: 984).

The finding began to make sense when gene expression profiling became available in the mid 1990s.

Gene expression profiling is done on a small square of glass or plastic, about 5 centimetres to a side. The square is called a **DNA microarray**, or a "gene chip." Short sequences of DNA, corresponding to known genes, are spotted onto the microarray in a grid pattern, so that investigators can localize and identify individual DNA sequences that bind. Next, a cell sample is taken, and its mRNA molecules are extracted. These represent the proteins that the cell is making. Because mRNA is unstable, researchers use an enzyme plus DNA nucleotides to copy the mRNA sequence into a complementary DNA or "cDNA" sequence. The set of cDNAs obtained from a cell, then, represents what the cell is doing—and not doing.

A gene expression profiling experiment is especially meaningful when it compares two sources of DNA. To tell the samples apart, the researcher incorporates a fluorescent molecule that emits light of a particular wavelength into the cDNAs from one patient and adds a different fluorescent molecule to DNA from a second source (Figure 6.2). For example, DNA from a cancerous white blood cell from a child who *did* respond to chemotherapy might be labeled red, and DNA from a child who *did not* respond to the same treatment might be labeled green. When both sources of cDNAs are added to the microarray and a laser is shone to activate the dyes, four colors result: for each spot on the grid, either red or green lights up if the embedded gene is expressed in one patient but not the other; yellow glows if both patients express the represented gene; and black results if neither patient is transcribing the probed gene. Intensity counts too. If a protein is very abundant, it fluoresces more brightly than a rare protein. However, abundance does not necessarily indicate import-ance. A computer algorithm prints a diagram of the microarray with DNA bound to it that looks like a checkerboard.

In 2001, researchers at the Dana-Farber Cancer Institute of Harvard University did gene expression profiles for children with acute leukemia, scanning 12,600 gene pieces embedded in micro-arrays. They found three distinctive patterns—not the two that they had expected for the long-recognized ALL and AML (Armstrong et al. 2002: 41). The new gene expression pattern corresponded to the patients whose chromosomes had switched parts, which was the same group whose cell surfaces had characteristics of both ALL and

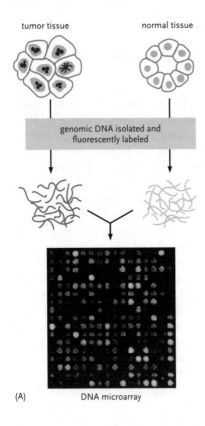

(A)   DNA microarray

Figure 6.2 DNA microarray. A DNA microarray is used to compare labeled DNA from two sources, such as tumor tissue and healthy tissue

Source: Adapted from *Molecular Biology of the Cell*, 5th edition, Figure 20–35A, p. 1239

AML. The investigators named the new type of cancer "mixed lineage leukemia" (MLL), and it accounted for the cases with a "very dismal prognosis."

The different types of evidence converged to describe the third type of acute leukemia at the gene, chromosome, and cellular levels. The MLL cells underexpressed about 1,000 genes and overexpressed about 200 genes. A critical gene, *MLL*, lies at the point on chromosome 11 where the chromosome swap occurs, in an event called a

translocation. The protein that the *MLL* gene encodes functions as a **transcription factor**, which is a protein that binds to thousands of genes, activating them to be transcribed into mRNA. When the *MLL* gene is moved to a new chromosomal address, the transcription factor behaves differently. It becomes an oncogene, activating a transcription program that speeds cell division rate. Cancer begins.

The nature of the genes whose expression differs in MLL compared with the other two leukemias illuminates how the cancer develops. All blood cells descend from very primitive stem cells in the bone marrow that give rise to two cell lineages. One, the myeloid lineage, gives rise to red blood cells, platelets, macrophages, and white blood cells called granulocytes. The other, the lymphoid lineage, gives rise to the T and B cells that carry out the immune response. AML begins after the myeloid lineage diverges from the initial stem cell, and ALL begins in the lymphoid lineage. In contrast, MLL affects a stem cell much earlier in the development of white blood cells, and so gives rise to surface features of *both* myeloid and lymphoid lineages—hence, the term *mixed lineage leukemia*.

DISTINGUISHING BREAST AND BRAIN CANCER SUBTYPES

Doctors and patients are already using gene expression profiling in selecting treatments for some forms of breast cancer. Oncotype DX is a test that looks at the expression of 21 genes in patients with early-stage breast cancer that has not spread to the lymph nodes and whose cells have receptors for the hormone estrogen. Tumor samples are given a score between 0 and 100, and patients with low numbers are considered less likely to benefit from chemotherapy (standard drugs and tamoxifen). More than 120,000 women have used the test. Doctors report that it has spared about a third of their patients from expensive, painful chemotherapy that would have been unlikely to have worked (Lo et al. 2010).

Gene expression profiling has limitations. It is helpful to patients only if the information is specific enough to guide treatment decisions. For example, MammaPrint is a test that looks at expression of 70 genes in breast cancer (Van't Veer et al. 2002: 530). MammaPrint was developed at the Netherlands Cancer Institute, where researchers examined preserved tumor samples going back five years, looking at all known genes. They identified 70 genes whose expression patterns differed between patients whose cancers spread and those whose cancers did not spread. Although this study is regarded as a "gold standard" in gene expression profiling experiments, one researcher points out that it doesn't extend what can be learned by assessing a cancer's division rate using other techniques (Koscielny 2010: 1). Even though MammaPrint looked at more genes than Oncotype DX, it did not analyze responses to specific treatments.

Interestingly, Oncotype DX and MammaPrint test for the expression of only one gene in common. However, they each look at genes involved in any of three pathways: cell division rate, presence or absence of estrogen receptors, and presence or absence of a growth factor receptor called HER2. This means that gene expression profiling probably needs to be more standardized before it becomes widely used in cancer care.

Gene expression profiling is also being used in brain cancer. The approach has identified four subtypes of glioblastoma multiforme, the most common form of brain cancer (Verhaak 2010). This cancer is usually fatal within 14 months of diagnosis. The nature of the genes whose expression is altered suggests that each of the four subtypes begins in a different type of cell, which may enable researchers better to target potential treatments. The subtyping of this brain cancer came from a project at the NIH in the US called the Cancer Genome Atlas, which is describing cancers in terms of their gene expression, rather than by body part or cell type. This information is useful in two general ways. Discovering that two cancers that affect different parts of the body actually are similar on a molecular level suggests that a treatment, perhaps an existing one, that works on one of the cancers might work on the other. Conversely, gene expression profiling can reveal when cancers of the same body part or cell type are actually *different* at the molecular level, perhaps explaining differences in treatment responses. This was the case for the three types of leukemia discussed earlier in the chapter.

The Cancer Genome Atlas is but one example of the new, genome-wide view of the human body, in health and disease. The next and final section of this final chapter looks at the abundance of information that is pouring forth from studies that consider entire genomes beyond our own.

A GENOMIC VIEW OF THE WORLD

Our view of the living world has traditionally reflected what we can see. This is a very limited sampling of biodiversity! A new field called **metagenomics** identifies and catalogs the organisms in a particular environment by collecting the DNA there, comparing the sequences to known sequences, and deducing the types of organism present, even those we cannot see or cultivate. Any area on Earth is a potential natural laboratory for metagenomic analysis, from the populations of bacteria and fungi that live under a toenail, to an entire ocean. In a practical sense, metagenomics may lead to discovery of natural products that are useful as drugs or energy sources. In a more philosophical sense, metagenomics is showing us that our technologies do not even begin to capture the spectrum of species that call planet Earth home.

THE HUMAN MICROBIOME

About 90 percent of the cells in our bodies are not actually our own— they are the bacteria that live in and on us. Only a few types of bacterium harm our health. US molecular biologist Joshua Lederberg coined the term "microbiome" to describe the microbial community of the human body (Lederberg and McCray 2001: 15). The NIH officially began the Human Microbiome Project in 2005, and, in 2007, the International Human Microbiome Consortium began, with members from Canada, Europe, Japan, the US, Australia, and Gambia.

All people share a core set of common bacteria, plus others from our personal environments. In one of the first human microbiome studies, researchers at Stanford University examined the feces in 14 babies' diapers at 26 times during their first year of life (Palmer et al. 2007: e177). The experiments showed that a newborn's intestines are clean—devoid of bacteria. But, within just a few weeks, each baby had acquired a distinctive "gut microbiome." However, by the first birthday, the infants had become much more similar in their intestinal inhabitants. An ongoing project is evaluating the microbiomes of 250 healthy adults from different populations to assess the variability of their bacterial passengers.

Metagenomics studies show that the human microbiome is densest in the mouth, intestines, nose, vagina, and skin. The numbers and diversity of our resident bacteria are staggering: 10 trillion bacteria live in our digestive systems. About 500 species of bacteria inhabit our mouths, and 350 of them cannot grow in a laboratory. When the Human Microbiome Project began, researchers estimated biodiversity by comparing part of just one very variable gene in all the DNA recovered from an area or body. Today, human microbiome experiments sequence all of the DNA.

FROM SEAS TO SUSHI TO BUG SPLATTER

Many metagenomics projects scrutinize habitats outside the human body. In 2007, researchers from the Institute for Biological Energy Alternatives in Rockville, MD, led by human genome pioneer J. Craig Venter, sequenced DNA from seawater samples taken from the Sargasso Sea (Venter 2004: 66). This 2 million-square-mile body of water off the coast of Bermuda is the location of the famed Bermuda Triangle, where navigators easily become lost. From more than 1 billion DNA bases sampled, the researchers identified 1.2 million new genes, representing 1,800 species, including 148 that were unknown. The researchers conducted a similar analysis of air samples taken from New York City.

Metagenomics meets forensics in a plate of sushi. This is not exactly a natural ecological community, but it does represent DNA from several species, although they may not be the species listed on the menu. In 2008, New York City teenagers Kate Stoeckle and Louisa Strauss applied a technique called DNA barcoding to sushi samples from ten grocery stores and four restaurants, to learn exactly which species were being served. The International Consortium for the Barcode of Life provided 30,562 genes that represent more than 5,000 species of fish. The young women found that a quarter of the samples from the restaurants and groceries were incorrectly labeled, with expensive types of sushi actually coming from inexpensive fish species.

Mislabeling food sources is a more serious concern than providing data for a science project or exposing a restaurant deceiving diners. A year after the students' investigation, researchers from the American Museum of Natural History published a similar study on 68 samples of tuna sushi from 31 restaurants in two cities (Lowenstein et al. 2009: e7866). Nineteen of the thirty-one restaurants do not actually sell what their menus list. As the researchers concluded,

"A piece of tuna sushi has the potential to be an endangered species, a fraud, or a health hazard." This study found all three.

Yet another novel habitat to explore for biodiversity is "windshield splatter," the DNA residue from unfortunate insects that have smashed into a moving car (Kosakovsky Pond et al. 2009: 2144). Researchers compared sampled insect DNA from cars driven up the largely urban east coast of the US, from Pennsylvania to Connecticut, and from Maine to New Brunswick, Canada, a rural route. As expected, the insect populations differed significantly. A great limitation of the study was that the database with which DNA samples were compared was heavily weighted with insect species that cause disease—many species must have been missing.

THE HUMAN GENOME SEQUENCE REVISITED

A human genome is packed with information, but not all of it may be biologically meaningful, and not all of it is meaningful in the same way. Only about 1 percent of the 3.2 billion bases of the human genome encodes proteins, although this small part accounts for at least 85 percent of disease-causing mutations. The protein-encoding part of the genome is called the **exome**.

Much of the non-exome part of the human genome consists of repeated sequences between protein-encoding genes, as well as parts of gene sequences that are cut out of mRNA molecules before they are translated into protein. When these "non-coding" parts of the genome were initially discovered, a few researchers termed them "junk," because they did not fit the reigning paradigm of gene encoding mRNA encoding protein, an idea so entrenched that it is called the "central dogma." The researchers likely used the word "junk" in jest, but unfortunately, over the years, the news media have taken it literally.

Researchers have always realized that we can't conclude that a DNA sequence is junk simply because we do not know what it does. The non-exome, in fact, includes many DNA sequences that have functions other than encoding protein. DNA sequences *control* protein synthesis and encode the RNAs other than mRNAs, such as the tRNAs and rRNAs, that are part of protein synthesis. The genome also encodes a collection of very short RNA molecules called, appropriately, microRNAs. These bind to specific sets of mRNAs, controlling which transcripts are translated into protein.

Much of the human genome isn't actually *human*, but instead consists of remnants of viruses that have inserted their DNA into our cells over the millennia, sometimes making us sick, but more often traveling silently from generation to generation along with our own genes. Genomes are also riddled with "copy number variants," which are short DNA sequences that are repeated next to each other, sometimes many times. The number of repeats may impart a different form of information than the protein-encoding DNA sequences of the exome.

GENE MAPS EVOLVE

DNA sequencing was invented in the 1970s. The two original techniques work by cutting many copies of the same DNA sequence into pieces that differ by one end base. Each of the four types of DNA base is labeled with a dye so that they can be distinguished. When the cut DNA pieces are aligned in size order, the overhanging bases are identified by their tags and assembled into a sequence, such as the italicized end base in the set of hypothetical DNA sequences in Table 6.2.

Back in the 1970s, a graduate student might have spent four or more years sequencing a single gene, as a doctoral project. Today, a student might use newer DNA sequencing methods, based on

---

Table 6.2 The concept behind traditional DNA sequencing

A set of copies of the same DNA sequence are aligned, and the end base is labeled:

ACTGTGCA*G*
ACTGTGC*A*
ACTGTG*C*
ACTGT*G*
ACTG*T*
ACT*G*
AC*T*
A*C*
A

Reading the end bases from the bottom reveals the sequence: *ACTGTGCAG*

---

nanotechnology and microfluidics, and sequence entire genomes of microbes in just a few days.

In the 1980s, researchers assembled genome maps that highlighted certain parts of the genome, called genetic markers. The maps grew ever denser. Each October, *Science* magazine published increasingly crowded chromosome depictions. That work did not cease when the first draft human genome sequence was published in 2001, because by then researchers had already shifted their focus to cataloging *variation* in the human genome. The HapMap project, an international effort, peppered human genome maps with SNPs, which became the raw material for the genome-wide association studies that are so common today.

Sequencing an entire 3.2 billion base human genome to learn one's health risks is a little like reading all of Wikipedia to learn about centipedes, typhoons, and refrigerators—it is simply too much information. Genome sequencing can reveal mutations that either cause disease or raise the risk of developing a particular disease, as well as recessive alleles that would never affect health unless a person had two copies of them. Testing for those particular genes may make more sense. Still, a few hundred people have had their genomes sequenced so far.

PERSONAL GENOME SEQUENCING

The first two individuals to have their genomes sequenced were, not surprisingly, scientist–celebrities J. Craig Venter, who led one of the human genome projects, and James Watson, co-discoverer of DNA's structure (Levy et al. 2007: e254; Wheeler et al. 2008: 872). Their results pointed out the limited value of knowing one's genome sequence. Venter's genome sequence revealed that he has blue eyes, lactose intolerance, dry earwax, novelty-seeking behavior, substance abuse, antisocial behavior, and a fondness for staying out late—all traits of which he and many others were already well aware. He learned that his body processes caffeine quickly. Said he at a genetics conference, "I can have two double lattes and wash it down with a Red Bull and not be affected by it." He also learned that he has an increased risk for heart disease and Alzheimer's disease, but he already knew this from his family history. Watson's genome sequence revealed that he carries several rare recessive disease alleles, but yielded few, if any, useful insights. Venter, who has obviously inherited a sense of humor genotype, joked that their genomes revealed that he and Watson are both bald, white scientists.

Despite the limited value of personal genome sequencing, several journalists, bloggers, and celebrities have had their genomes "done," like the latest fashion fad. Projects have proliferated. First came the Personal Genome Project, to sequence the genomes of ten well-known individuals, at Harvard University. The "1,000 Genomes Project" and a "20,000 Genomes Project" were swiftly announced. Meanwhile, those in the genomic know blogged and wrote articles about what they'd learned—which generally wasn't much beyond what they could have learned from their family trees.

When is sequencing one's genome actually useful? Ironically, for one group of investigators, the research went full circle back to the single gene level. Richard Lifton and colleagues at the Yale School of Medicine developed a DNA microarray chip that bears the 34 million DNA bases that constitute the human exome. They tested the value of the exome array with a sample from an infant in Turkey who had severe diarrhea and failure to gain weight. The symptoms were similar to those of a disease called Bartter syndrome, but the child's doctors weren't convinced. Applying the baby's DNA to the exome array revealed, within just hours, that the baby had a condition called congenital chloride diarrhea. Lifton thinks the cost of the whole-exome approach will one day be low enough to become routine in solving tough diagnostic cases, like that of the baby in Turkey.

Personal genome sequencing is likely to be the most useful on a much larger scale than considering a single gene in a single person. Every few months, genetics journals publish the genome sequence of a person from a particular population group, in an ongoing effort to catalog and understand human genetic diversity. Chapter 1 described the genome sequencing of a modern day Bantu and a Khoisan.

Other reasons to sequence genomes, of humans and others, include:

- to learn how species are related through evolution;
- to track the changes that occur in a genome as cancer, or an infectious disease, or a birth defect progresses;
- to obtain information on susceptibilities and drug reactions to guide health care and lifestyle choices;

- to reconstruct human evolution and migrations;
- to identify genetic contributions to such hard-to-define traits as intelligence, personality, mood, behaviors, and beauty;
- to search the genomes of centenarians—people who live past 100—to discover longevity genes.

It will also be important, as personal genome sequencing becomes more accessible, to track how people use the information in their genomes.

EPIGENETICS

When the idea to sequence the human genome was just a dream, one of the arguments against doing so was that it didn't seem useful to do it just because we could. This was dubbed the "Mount Everest" argument, sequencing the genome simply because it is there, like climbing the mountain. The argument remains valid, because a genome does not function in isolation, and knowing the long sequences of A, C, T, and G may not matter much. Making sense of a genome sequence depends, to an extent, on changes beyond the DNA sequence that affect how genes are used, termed **epigenetics**. These changes occur as signals from inside and outside the cell affect which sets of certain chemical groups actually contact exposed DNA, enabling it to be transcribed or blocking it from being transcribed.

In cell nuclei, certain types of chemical groups jump on and off the DNA, in tune to various signals. One type of epigenetic change is to move the histone proteins around which DNA coils, which exposes certain genes to RNA polymerase, to be transcribed, and shields others. Another type of epigenetic change is to fasten methyl (CH_3) groups to certain cytosines, the "C"s in the DNA four-letter alphabet. For example, the pattern of methylation in a human embryonic stem cell, which has not specialized, is different from that of a fibroblast, which, as a connective tissue cell, specializes in secreting collagen (Lister et al. 2009: 315). Understanding the layers of epigenetic controls that are painted over our genomes will enhance our utilization of human genome information.

When mathematical calculators came on the market at just about the time when DNA sequencing was invented, in the 1970s, they cost more than $100. Today, they are very inexpensive. Personal genome sequencing may follow the same path, and, by the time that you finish reading this book, having your genome sequenced for a reasonable price may be close at hand. Will you do it? What would you like to learn? How will you use the information?

It is a lot to consider!

WHAT WOULD YOU DO?

Our DNA is like a biological crystal ball in that it can foretell the future. Also like a crystal ball, our DNA sequences cannot account for what we experience as we go through life. Genes and the environment mold who we are and how we live, and these two forces interact in complex ways that we are only beginning to fathom. Therefore, genetic tests can provide valuable information, but that information does not function alone. We can do a great deal to temper even a dire diagnosis.

Genetic tests are possible at many points in human development and the life cycle. At which of the following times might you consider taking a genetic test—and why?

- preimplantation genetic diagnosis to choose the genome of your child (see What Would You Do? in Chapter 1);
- a prenatal test to detect a genetic disease in an embryo or fetus;
- at birth, as part of a newborn screening panel to detect inborn errors of metabolism and other treatable, single-gene diseases;
- as a teen, to detect a genotype for an adult-onset condition such as Huntington's disease;
- a battery of direct-to-consumer DNA tests, the interpretation of which is based on "associations" that may be disproven as data accumulate;
- as a young adult about to marry, to learn which single-gene diseases you and your partner carry;
- as an older adult, to learn how you might die.

Taking a genetic test requires a great deal of careful thought.

SUGGESTED READING

Several books chronicle the sequencing of the human genome, as well as the research trajectories that have diverged from knowing the sequence. Kevin Davies tells of the genome project in *Cracking the Genome: Inside the Race to Unlock Human DNA* (Johns Hopkins University Press, 2002) and of the commercialization of the process in *The $1,000 Genome* (Free Press, Simon & Schuster, 2011). The two leaders of the human genome projects have written books. Francis S. Collins wrote *Language of Life: DNA and the Revolution in Personalized Medicine* (HarperCollins, 2010), and J. Craig Venter tells of his own genome sequencing in *A Life Decoded: My Genome: My Life* (Viking Adult, 2007). A look at genomes other than our own is in *The New Science of Metagenomics: Revealing the Secrets of Our Microbial Planet* (Washington, DC: National Academy Press, 2007). Perhaps you will one day write about your own genome!

GLOSSARY OF TERMS

Adenine: One of the four nitrogenous bases in DNA. Complementary base pairs to thymine (T).

Allele: A variant of a gene.

Amino acid: A protein building block.

Antibodies: Proteins produced in B cells and built of Y-shaped subunits that bind non-self antigens, signaling other parts of the immune response.

Antigen: A molecule, or part of one, that stimulates an immune response.

Autosomal dominant: Mode of inheritance in which an allele on a non-sex chromosome that causes the phenotype is present in one copy. May affect either sex.

Autosomal recessive: Mode of inheritance in which an allele on a non-sex chromosome that causes the phenotype must be present in two copies. May affect either sex.

Autosome: A chromosome that does not include a gene that determines sex.

Balancing selection: Maintenance of a mutant allele in a population because the carrier state has a survival or reproductive advantage over individuals who do not have the mutant allele.

Bioinformatics: Computer technology that analyzes DNA sequence data.

Cell: The structural and functional unit of a living organism.

Cell cycle: The sequence of events that a cell undergoes, including division (mitosis) and the time spent in preparation for division (interphase).

Chromosome: A structure in a cell's nucleus, built of DNA and associated proteins, that contains the genes and appears rod-like during cell division. A human cell has 23 pairs of chromosomes.

Cline: An allele frequency that increases or decreases across a geographical area.

Cytokines: Protein products of T cells that have protective functions, including signaling bone marrow to manufacture lymphocytes, detecting virally infected cells, causing fever, and destroying bacterial toxins.

Cytoplasm: Gel-like substance in a cell, outside the nucleus, in which organelles are suspended or move.

Cytosine: One of the four nitrogenous bases in DNA. Complementary base pairs to guanine (G).

Dihybrid cross: Breeding individuals who are heterozygous for the same two genes.

Diploid: A cell that has two copies of each chromosome type.

DNA: The genetic material, deoxyribonucleic acid. DNA is a long molecule built of sequences of four nitrogen-containing bases: adenine (A), guanine (G), cytosine (C), and thymine (T). The bases are held in pairs by two backbone structures, forming a double helix.

DNA microarray: A small glass or plastic chip that bears short DNA sequences which serve as probes for complementary DNA sequences in patient samples.

Dominant: An allele that affects a phenotype in one copy.

Epigenetics: Changes to DNA that affect its expression, other than changing the base sequence.

Eugenics: Attempts to alter the future gene pool.

Exome: The part of the genome that encodes protein.

Expressivity: Variability in symptom severity among people with the same genotype.

Gene: A DNA sequence that encodes a protein's amino acid sequence.

Gene expression profiling: A technique that identifies and measures the abundance of the set of mRNA molecules present in a particular cell type under specific conditions, providing a snapshot of the cell's activities.

Gene pool: All of the alleles in a population.

Genetic ancestry: Comparing DNA sequences to infer familial relationships.

Genetic determinism: The idea that our genes set all of our traits, and other factors, such as the environment, cannot alter them.

Genetic drift: A subpopulation sequesters certain allele combinations from the larger group.

Genome: One set of the DNA information in a cell. A genome is species-specific and also varies among individuals of the same species to a small extent.

Genome-wide association study: A comparison of many sites in the genome between two groups of individuals, who differ in an important way, but are matched in others. Used to identify genes that contribute to the risk of developing a particular condition.

Genotype: The specific alleles for a gene or genes in an individual.

Guanine: One of the four nitrogenous bases in DNA. Complementary base pairs to cytosine (C).

Haplogroup: Marker sets on the Y chromosome or mitochondrial DNA, used to trace ancestry.

Haploid: A cell with a single set of chromosomes; egg or sperm.

Haplotype: The alleles on a segment of a chromosome, which tend to be inherited together as a unit.

Heterozygote: An individual with two different alleles of a gene.

Hominin: An individual on the branch of the evolutionary tree that gave rise to humans only. Also called hominid.

Homozygote: An individual with two identical alleles of a gene.

Independent assortment: The inheritance pattern of a gene on one chromosome does not affect the inheritance of a gene on another chromosome.

Lymphocyte: A type of white blood cell; a T or B cell.

Meiosis: Form of cell division that gives rise to sex cells (sperm or egg), halving their chromosome number.

Messenger RNA: Type of RNA that includes the information for a specific amino acid sequence.

Metagenomics: Identifying the organisms that live in a particular environment by sampling and sequencing their DNA.

Mitosis: Form of cell division that maintains the chromosome number in somatic (body) cells.

Monohybrid cross: Breeding two individuals who are heterozygous for the same gene.

Mutant: An unusual or abnormal expression of a gene.

Mutation: A change in a DNA sequence, natural or induced.

Natural selection: The differential survival of individuals whose inherited trait variants better adapt them to reproduce successfully in a particular environment.

Nucleotide: A DNA building block, consisting of a phosphate group, a sugar, and a nitrogenous base.

Nucleus: The genetic headquarters of a cell, which encloses DNA.

Oncogene: A gene that causes cancer when overexpressed.

Organelles: Structures in cells that carry out specific functions.

Penetrance: The percent of individuals with the same genotype who express the associated phenotype.

Phenotype: The expression of an allele combination (genotype).

Pleiotropy: The association of several symptoms with a genotype.

Protein: A long molecule built of amino acids that provides or contributes to an inherited trait.

Recessive: An allele that affects a phenotype when present in two copies.

Ribosomal RNA: A component of a ribosome.

Ribosome: A structure in a cell built of RNA and protein molecules on which translation of mRNA into protein occurs.

RNA: Ribonucleic acid; an informational molecule that is an intermediate language between DNA and protein.

Segregation: Separation of two copies of a gene into separate gametes.

Sex chromosome: A chromosome that includes a gene that determines sex.

Single nucleotide polymorphism (SNP): A single-base site in the genome that differs in more than 1 percent of a population.

Thymine: One of the four nitrogenous bases in DNA. Complementary base pairs to adenine (A).

Telomere: A chromosome tip.

Transcription: RNA synthesis.

Transcription factor: A protein that turns on specific groups of genes.

Transfer RNA:Type of RNA that brings a specific amino acid to a particular mRNA codon (three-base sequence).

Translation: Protein synthesis.

Translocation: An event in which chromosomes exchange parts, or one chromosome attaches to another.

Tumor suppressor: A gene that, when deleted or inactivated, causes cancer.

Wild type: The most common allele in a population; the "normal" allele.

X-linked recessive: Mode of inheritance in which a male expresses a trait whose gene is on the X chromosome because he does not have a second allele to mask it; a female requires two copies of the allele to express the trait.

BIBLIOGRAPHY

Aiuti, A., Slavin, S., Aker, M., et al. (2002) "Correction of ADA-SCID by stem cell gene therapy combined with nonmyeloablative conditioning," *Science*, 296: 2410–2413.

Amado, D., Mingozzi, F., Hui, D., et al. (2010) "Safety and efficacy of subretinal readministration of a viral vector in large animals to treat congenital blindness," *Science Translational Medicine*, 2: 1–9.

Andersen, D.H. (1944) "Celiac syndrome III," *American Journal of Diseases of Children*, 70: 100–113.

Armstrong, S.A., Staunton, J.E., Silverman, L.B., et al. (2002) "*MLL* translocations specify a distinct gene expression profile that distinguishes a unique leukemia," *Nature Genetics*, 30: 41–47.

Asfaw, B., Gilbert, W.H., Beyene, Y., et al. (2002) "Remains of *Homo erectus* from Bouri, Middle Awash, Ethiopia," *Nature*, 416: 317–320.

Avery, O., MacLeod, C., and McCarty, M. (1944) "Studies on the chemical nature of the substance inducing transformation of pneumococcal types," *Journal of Experimental Medicine*, 79: 137–158.

Beadle, G.W. and Tatum, E.L. (1941) "The genetic control of biochemical reactions in *Neurospora*," *Proceedings of the National Academy of Sciences*, 27: 499–506.

Behar, D.M., Villems, R., and Soodyall, H. (2008) "The dawn of human matrilineal diversity," *American Journal of Human Genetics*, 82: 1130–1140.

Beutler, E., Felitti, V.J., and Koziol, J.A. (2002) "Penetrance of 845 G-A (C282Y) HFE hereditary hemochromatosis mutation in the U.S.," *Lancet*, 359: 210–218.

Bickel H., Gerrard J., and Hickmans E.M. (1953) "Influence of phenylalanine intake on phenylketonuria," *Lancet*, 2: 812–813.

Boyle, M.P. (2007) "Adult cystic fibrosis," *Journal of the American Medical Association*, 298: 1787–1793.

Brown, M.S. and Goldstein, J.L. (1976) "Receptor-mediated control of cholesterol metabolism," *Science*, 191: 150–154.

Buck, P.S. (1951) *The Child Who Never Grew*, Bethesda, MD: Woodbine House.

Cann, R.L., Stoneking, M., and Wilson, A.C. (1987) "Mitochondrial DNA and human evolution," *Nature*, 325: 31–36.

Cartier, N., Hacein-Bey-Abina, S., Bartholomae, C.C., et al. (2009) "Hematopoietic stem cell gene therapy with a lentiviral vector in X-linked adrenoleukodystrophy," *Science*, 326: 818–823.

Castellani, C., Picci, L., Tamanini, A., et al. (2009) "Association between carrier screening and incidence of cystic fibrosis," *Journal of the American Medical Association*, 302: 2573–2579.

Cavazzana-Calvo, M., Hacein-Bey, S., Basile, G.S., et al. (2000) "Gene therapy of human severe combined immunodeficiency (SCID)-X1 disease," *Science*, 288: 669–672.

Cohen, J. (2009) "HIV natural resistance field finally overcomes resistance," *Science*, 326: 1476–1477.

Collins, F.S. (2010) *Language of Life: DNA and the Revolution in Personalized Medicine*, New York: HarperCollins.

Crick, F. (1966) *Of Molecules and Men*, Seattle and London: University of Washington Press.

Crick, F.H., Barnett, L., Brenner, S., and Watts-Tobin, R.J. (1961) "General nature of the genetic code for proteins," *Nature*, 192: 1227–1232.

Dahl, M., Tybjaerg-Hansen, A., Lange, P., and Nordestgaard, B.G. (2005) "DeltaF508 heterozygosity in cystic fibrosis and susceptibility to asthma," *Lancet*, 351: 1911–1913.

Davies, K. (2002) *Cracking the Genome: Inside the Race to Unlock Human DNA*, Baltimore and London: Johns Hopkins University Press.

Davies, K. (2011) *The $1,000 Genome*, New York: Free Press, Simon and Schuster.

di Sant'Agnese, P.A., Darling R.C., Perera G.A., and Shea, E. (1953) "Abnormal electrolyte composition of sweat in cystic fibrosis of the pancreas," *Pediatrics*, 12: 549–563.

Dormer, R.L., Harris, C.M., Clark, Z., et al. (2005) "Sildenafil (Viagra) corrects DeltaF508-CFTR location in nasal epithelial cells from patients with cystic fibrosis," *Thorax*, 60: 55–59.

Farrell, P., LeMarechal, C., Ferec, C., et al. (2007) "Discovery of the principal cystic fibrosis mutation (*F508del*) in ancient DNA from Iron Age

Europeans," Nature Precedings (http://hdl.handle.net/10101/npre.2007. 1276.1).

Feder, J., Gnirke, A., Thomas, W., et al. (1996) "A novel MHC class 1-like gene is mutated in patients with hereditary haemochromatosis," *Nature Genetics*, 13: 399–408.

Fölling, A. (1934) "On excretion of phenylpyruvic acid in the urine as an anomaly of metabolism in connection with mental retardation," *Hoppe Seylers Zeitschrift für Physiologische Chemie*, 227: 169–176.

Fölling, I. (2008) "The discovery of phenylketonuria," *Acta Paediatrica*, 83: 4–10.

Fost, N. and Kaback, M.M. (1973) "Why do sickle screening in children?" *Pediatrics*, 51: 742–745.

Franklin, R. and Gosling, R.G. (1953) "Molecular configuration in sodium thymonucleate," *Nature*, 171: 740–741.

Garrod, A. (1902) "The incidence of alkaptonuria: a study in chemical individuality," *Lancet*, 11: 1616–1620.

Gelsinger, P. (2001) "Jesse's intent," available online at www.circare.org/submit/jintent.pdf.

Goebel, T., Waters, M.R., O'Rourke, D.H., et al. (2008) "The late Pleistocene dispersal of modern humans in the Americas," *Science*, 319: 1497–1502.

Gracey, M. and King, M. (2009) "Indigenous health part 1: Determinants and disease patterns," *Lancet*, 374: 65–75.

Griffith, F. (1928) "The significance of pneumococcal types," *Journal of Hygiene*, 27: 113–159.

Gross, S.J., Bletcher, F.A., Monoghan, K.G. (2008) "Carrier screening in individuals of Ashkenazi Jewish descent," *Genetics in Medicine*, 10: 54–57.

Gusella, J.F., Wexler, N.S., Conneally, P.M., et al. (1983) "A polymorphic DNA marker genetically linked to Huntington's disease," *Nature*, 306: 234–238.

Guthrie, R. and Susi, A. (1963) "A simple phenylalanine method for detecting phenylketonuria in large populations of newborn infants," *Pediatrics*, 32: 338–343.

Guthrie, R. and Whitney, S. (1964) "Phenylketonuria detection in the newborn infant as a routine hospital procedure. A trial of the phenylalanine screening method in 400,000 infants," *Children's Bureau Pub.* No. 419. US Government Print Office, Washington DC.

Hale, J.E., Parad, R.B., and Comeau, A.M. (2008) "Newborn screening showing decreasing incidence of cystic fibrosis," *New England Journal of Medicine*, 358: 973–974.

Hall, S.S. (2008) "Last of the Neanderthals," *National Geographic*, October 2008: 34–59.

Haller, J.O., Berdon, W.E., and Franke, H. (2001) "Sickle cell anemia: The legacy of the patient (Walter Clement Noel), the intern (Ernest Irons), and the attending physician (James Herrick) and the facts of its discovery," *Pediatric Radiology*, 31: 889–890.

Hampton, M.L. (1974) "Sickle cell 'nondisease'," *American Journal of Diseases of Children*, 128: 58–61.

Harper, P.S., Lim, C., and Craufurd, D. (2000) "Ten years of presymptomatic testing for Huntington's disease: The experience of the UK Huntington's Disease Prediction Consortium," *Journal of Medical Genetics*, 37: 567–571.

Henig, R.M. (2000) *The Monk in the Garden*, New York: Houghton Mifflin Co.

Herrick, J.B. (1910) "Peculiar elongated and sickle-shaped red blood corpuscles in a case of severe anemia," *Archives of Internal Medicine*, 6: 517–521.

Hershey, A.D. and Chase, M. (1952) "Independent functions of viral protein and nucleic acid in growth of bacteriophage," *Journal of General Physiology*, 36: 39–56.

Hoagland, M. (1990) *Toward the Habit of Truth*, New York: W.W. Norton & Co.

Hollenberg, M.D., Kaback, M.M., and Kazazian, H.H. (1971) "Adult hemoglobin synthesis by reticulocytes from the human fetus at midtrimester," *Science*, 174: 698–702.

Huntington G. (1872) "On chorea," *Medical Surgical Reporter*, 26: 317–321.

Huntington's Disease Collaborative Research Group (1993) "A novel gene containing a tri-nucleotide repeat that is expanded and unstable on Huntington's disease chromosomes," *Cell*, 2: 971–983.

Hutter, G. (2009) "Long-term control of HIV by CCR5 delta32/delta32 stem cell transplantation," *New England Journal of Medicine*, 360: 692–698.

International Human Genome Sequencing Consortium (2001) "Initial sequencing and analysis of the human genome," *Nature*, 409: 860–921.

Jervis, G. (1937) "Phenylpyruvic oligophrenia: Introductory study of 50 cases of mental deficiency associated with excretion of phenylpyruvic acid," *Archives of Neurology and Psychiatry*, 38: 944.

Johanson, D. and White, T. (1979) "A systematic assessment of early African hominids," *Science*, 202: 321–330.

Jungers, W.L. (2010) "Biomechanics: Barefoot running strikes back," *Nature*, 463: 433–434.

Kaback, M.M., Lim-Steele, J., Dabholkar, D., et al. (1993) "Tay–Sachs disease—carrier screening, prenatal diagnosis, and the molecular era. An international perspective, 1970 to 1993. The International TSD Data

Collection Network," *Journal of the American Medical Association*, 270: 2307–2315.

Kan, Y.W. and Dozy, A.M. (1978) "Antenatal diagnosis of sickle-cell anemia by D.N.A. analysis of amniotic-fluid cells," *Lancet*, 8096: 910–912.

Karafet, T.M., Mendez, F.L., and Mellerman, M.B. (2008) "New binary polymorphisms reshape and increase resolution of the human Y chromosomal haplogroup tree," *Genome Research*, 18: 830–838.

Ke, Y., Su, B., Song, X., et al. (2001) "African origin of modern humans in East Asia: A tale of 12,000 Y chromosomes," *Science*, 292: 1151–1153.

Keller, E.F. (2000) *The Century of the Gene*, Cambridge, MA, and London: Harvard University Press.

Kerem, B., Rommens, J.M., Buchanan, J.A., et al. (1989) "Identification of the cystic fibrosis gene: genetic analysis," *Science*, 245: 1073–1080.

Kessler, W.R. and Andersen, D.H. (1951) "Heat prostration in fibrocystic disease of the pancreas and other conditions," *Pediatrics*, 8: 648–655.

Keys, A. (1980) *Seven Countries: A Multivariate Analysis of Death and Coronary Heart Disease*, Cambridge, MA: Harvard University Press.

Knowles, M., Gatzy, J., and Boucher, F.C. (1981) "Increased bioelectric potential difference across respiratory epithelia in cystic fibrosis," *New England Journal of Medicine*, 305: 1489–1495.

Knowlton, R.G., Cohen-Haguenauer, O., Van Cong, N., et al. (1985) "A polymorphic DNA marker linked to cystic fibrosis is located on chromosome 7," *Nature*, 318: 380–382.

Kosakovsky Pond, S., Wadhawan, S., and Chlaromonte, F. (2009) "Windshield splatter analysis with the Galaxy metagenomic pipeline," *Genome Research*, 19: 2144–2153.

Koscielny, S. (2010) "Why most gene expression signatures of tumors have not been useful in the clinic," *Science Translational Medicine*, 2: 1–3.

Laleuza-Fox, C., Rompler, H., Caramelli, D., et al. (2007) "A melanocortin 1 receptor allele suggests varying pigmentation among Neanderthals," *Science*, 318: 1453–1455.

Lederberg, J. and McCray, A.T. (2001) " 'Ome Sweet' Omics—a genealogical treasury of words," *The Scientist*, 15: 8.

Levy, S., Sutton, G., Ng, P.C., et al. (2007) "The diploid genome sequence of an individual human," *PLoS Biology*, 5: e254.

Lewis, R. (1993) "Gene discovery: The giant first step toward therapy," *The Scientist*, 7(13): 14.

Li, J.Z., Absher, D.M., Tang, H., et al. (2008) "Worldwide human relationships inferred from genome-wide patterns of variation," *Science*, 319: 1100–1104.

Liou, T.G. and Rubenstein, R.L. (2009) "Carrier screening, incidence of cystic fibrosis, and difficult decisions," *Journal of the American Medical Association*, 302: 2595–2596.

Lister, R., Pelizzola, M., Down, R.H., et al. (2009) "Human DNA methylomes at base resolution show widespread epigenomic differencs," *Nature*, 462: 315–322.

Liu, H., Prugnolle, F., Manica, A., and Balloux, F. (2006) "A geographically explicit genetic model of worldwide human-settlement history," *American Journal of Human Genetics*, 79: 230–236.

Lo, S.S., Mumby, P.B., Norton, J., et al. (2010) "Prospective multicenter study of the impact of the 21-gene recurrence score assay on medical oncologist and patient adjuvant breast cancer treatment selection," *Journal of Clinical Oncology*, 28: 1671–1676.

Longinotti, M., Pistidda, P., Oggiano, L., et al . (2008) "A 12-year preventive program for beta-thalassemia in Northern Sardinia," *Clinical Genetics*, 46: 238–243.

Lowenstein, J.H., Amato G., Kolokotronis, S.-O. (2009) "The real *maccoyii*: Identifying tuna sushi with DNA barcodes—contrasting characteristic attributes and genetic distances," *PLoS ONE*, 4(11): e7866.

McCabe, L. (1996) "Efficacy of a targeted genetic screening program for adolescents," *American Journal of Human Genetics*, 59: 762–763.

McElheny, V.K. (2003) *Watson and DNA*, Cambridge, MA: Perseus Publishing.

McElheny, V.K. (2010) *Drawing the Map of Life: Inside the Human Genome Project*, New York: Basic Books.

Mendel, G. (1865) "Experiments on plant hybrids," *Verhandlungen des naturforschenden den vereines in Brunn*, 4: 3–47.

Meselson, M. and Stahl, F.W. (1958) "The replication of DNA in *Escherichia coli*," *Proceedings of the National Academy of Sciences*, 44: 671–682.

Miescher, F. (1871) "Ueber die chemische Zusammensetzung der Eiterzellen," *Medicinisch-chemische Untersuchunger*, 4: 441–460.

Mitchell, J.J., Capua, A., Clow, C., and Scriver, C.R. (1996) "Twenty-year outcome analysis of genetic screening programs for Tay–Sachs and beta-thalassemia disease carriers in high schools," *American Journal of Human Genetics*, 59: 793–798.

Ng, P.C., Murray, S.S., Levy, S., and Venter, J.C. (2009) "An agenda for personalized medicine," *Nature*, 461: 724–726.

O'Brien, J.S., Okada, S., Fillerup, D.L., et al. (1971) "Tay–Sachs disease: Prenatal diagnosis," *Science*, 172: 61–64.

Orel, V. (1996) *Gregor Mendel the First Geneticist*, Oxford, New York, Tokyo: Oxford University Press.

Painter, T.S. (1923) "Further observations on the sex chromosomes of mammals," *Science*, 28: 247–248.

Palmer, C., Bik, E.M., DiGiulio, D.B., et al. (2007) "Development of the human infant intestinal microbiota," *PLoS Biology*, 5: e177.

Pauling, L., Harvey, A.I., Singer, S.J., and Wells, I.C. (1949) "Sickle cell anemia, a molecular disease," *Science*, 110: 543–548.

Pier, G.B., Grout, M., Zaldi, T., et al. (1998) "*Salmonella typhi* uses CFTR to enter intestinal epithelial cells," *Nature*, 393: 79–82.

Poolman, E.M. and Galvani, A.P. (2007) "Evaluating candidate agents of selective pressure for cystic fibrosis," *Journal of the Royal Society Interface*, 4: 91–98.

Quinton, P.M. (2007) "Cystic fibrosis: Lessons from the sweat gland," *Physiology*, 22: 212–225.

Riordan, J.R., Rommens, J.M., Kerem, B., et al. (1989) "Identification of the cystic fibrosis gene: Cloning and characterization of complementary DNA," *Science*, 245: 1066–1072.

Saleheen, D. and Frossard, P.M. (2008) "The cradle of the deltaF508 mutation," *Journal of Ayub Medical College*, 20: 157–159.

Schuster, S.C., Miller W., Ratan, A., et al. (2010) "Complete Khoisan and Bantu genomes from southern Africa," *Nature*, 463: 943–947.

Scriver, J.B. and Waugh, T.R. (1930) "Studies on a case of sickle-cell anemia," *Canadian Medical Association*, 23: 375–380.

Shreeve, J. (2004) *The Genome War: How Craig Venter Tried to Capture the Code of Life and Save the World*, New York: Alfred A. Knopf.

Siperstein, M.D. and Fagan, V.M. (1966) "Feedback control of mevalonate synthesis by dietary cholesterol," *Journal of Biological Chemistry*, 241: 602–609.

Slany, R.K. (2009) "The molecular biology of mixed lineage leukemia," *Haematologica*, 94: 984–993.

Stedman, H.H., Kozyak, B.W., Nelson, A., et al. (2004) "Myosin gene mutation correlates with anatomical changes in the human lineage," *Nature*, 428: 415–419.

Stix, G. (2008) "Traces of a distant past," *Scientific American*, July 2008: 56–64.

Stone, R. (2010) "Leprosy's last stand—or early days of a war of attrition?" *Science*, 327: 939.

Sturtevant, A.H. (1913) "The linear arrangement of six sex-linked factors in *Drosophila*, as shown by their mode of association," *Journal of Experimental Zoology*, 14: 43–59.

Sturtevant, A.H. (1965) *A History of Genetics*, Cold Spring Harbor, NY: Cold Spring Harbor Press.

Tennet, D.M., Siegel, H., Zanetti, M.E., et al. (1960) "Plasma cholesterol lowering action of bile acid binding polymers in experimental animals," *Journal of Lipid Research*, 1: 469–473.

Tishkoff, S.A., Gonder, K., Hirbo, J., et al. (2003) "The genetic diversity of linguistically diverse Tasmanian populations: a multilocus analysis," *American Journal of Physical Anthropology Suppl.*, 36: 208–209.

Trousseau, A. (1865) *Clinique medicale de l'hotel de Paris*, vol. II, Paris: J.-B. Balliere.

Tsui, L.-C., Buchwalk, M., Barker, D., et al. (1985) "Cystic fibrosis locus defined by a genetically linked polymorphic DNA marker," *Science*, 230: 1054–1057.

Undenfriend, S. and Cooper, J.R. (1952) "The enzymatic conversion of phenylalanine to tyrosine," *Journal of Biological Chemistry*, 194: 503–511.

Van't Veer, L.J., Dai, H., van de Vijver, M.J., et al. (2002) "Gene expression profiling predicts clinical outcome of breast cancer," *Nature*, 415: 530–536.

Venter, J.C. (2007) *A Life Decoded: My Genome: My Life*, New York: Viking Adult.

Venter, J.C., Adams, M.D., Myers, E.W., et al. (2001) "The sequence of the human genome," *Science*, 291: 1304–1351.

Venter, J.C., Remington, K., Heidelberg, J.F., et al. (2004) "Environmental genome shotgun sequencing of the Sargasso Sea," *Science*, 304: 66–74.

Verhaak, R.G.W., Hoadley, K.A., Purdom, E., et al. (2010) "Integrated genomic analysis identifies clinically relevant subtypes of glioblastoma characterized by abnormalities in PDGFRA, IDH1, EGFR, and NF1," *Cancer Cell*, 9: 157–173.

Wang, S., Lewis, C.M., Jakobsson, M., et al. (2007) "Genetic variation and population structure in Native Americans," *PLoS Genetics*, 3(11): e185.

Watson, J.D. and Crick, F.H. (1953a) "Molecular structure of nucleic acids; a structure for deoxyribose nucleic acid," *Nature*, 171: 737–738.

Watson J.D. and Crick F.H. (1953b) "Genetical implications of the structure of deoxyribose nucleic acid," *Nature*, 171: 964.

Watson, J.D. (1968) *The Double Helix: A Personal Account of the Discovery of the Structure of DNA*, New York: Simon and Schuster,.

Watson, J.D. (2003) *DNA the Secret of Life*, New York: Alfred A. Knopf,.

Wells, H.G., Huxley, J.S., and Wells, G.P. (1929) *The Science of Life*, New York: The Literary Guild.

Wheeler, D.A., Srinirasan, M., Egholm, M., et al. (2008) "The complete genome of an individual by massively parallel DNA sequencing," *Nature*, 452: 872–876.

White, T., Asfaw, B., Beyene, Y., et al. (2009) "*Ardipithecus ramidus* and the paleobiology of early hominids," *Science*, 326: 75–86.

White, T., Asfaw, B., DeGusta, D., et al . (2003) "Pleistocene *Homo sapiens* from Middle Awash, Ethiopia," *Nature*, 423: 742–747.

Wilkins, M.H.F. (1953) "Molecular structure of deoxypentose nucleic acids," *Nature*, 171: 738–740.

Woo, S.L.C., Lidsky, A.S., Guttler, F., et al. (1983) "Cloned human phenylalanine hydroxylase gene allows prenatal diagnosis and carrier detection of classical phenylketonuria," *Nature*, 306: 151–155.

Zhang, F-R., Huang, W., Chen, S-M., et al. (2009) "Genome-wide association study of leprosy," *New England Journal of Medicine*, 361: 2609–2618.

INDEX